Intergenerational D
Environmental Justi
the Case of Nuclear

This book explores the interplay between intergenerational justice and intragenerational justice using nuclear waste management as a consistent case to explore these themes.

Lee Towers and Matthew Cotton examine the issue of intergenerational justice from a social scientific perspective, drawing on central case studies of nuclear waste management in Canada, Finland, and the United Kingdom. They connect indigenous philosophies and notions of justice with the concept of intergenerational democracy, advocating for better inclusion of youth and elders in decision-making that affects their well-being. As such, the book's primary objectives are fourfold:

- To assess whether trade-offs between intergenerational and intragenerational justice are necessary, and if so, what these trade-offs are and how they might be resolved.
- To critically assess dominant western liberal philosophical approaches that shape contemporary intergenerational justice thinking in policy and practice, and consider alternatives drawn from anthropology and indigenous philosophies.
- To assess how far our current capitalist system can achieve substantive forms of justice.
- To critically examine three nuclear waste management case studies and assess how far these achieve environmental and energy justice and how they exemplify tensions between inter- and intragenerational justice.

This short, accessible volume will be of great interest to students and scholars of energy, environmental justice, and ethics.

Lee Towers is a postdoctoral researcher working at Teesside University looking into aspects of intra- and intergenerational justice and nuclear waste solutions. He holds a PhD in applied social sciences from Brighton University. This PhD explored energy justice with a focus on community energy organisations and their work on energy poverty and climate mitigation in the United Kingdom. Previous published work includes an examination of community energy work on reducing energy poverty in the UK privatised energy system and an exploration of the intergenerational aspects of the pandemic published by Brighton University. His current postdoctoral position is funded by the Nuclear Waste Services.

Matthew Cotton is professor of environmental justice and public policy in the School of Social Sciences, Humanities, and Law at Teesside University. He holds a PhD in environmental science from the University of East Anglia. His research explores the social and ethical dimensions of technology development and environmental planning, and the effective involvement of stakeholders in questions of socio-economic and ecological justice. His previous published works on these topics include the monographs: *Virtual Reality, Empathy and Ethics*; *Nuclear Waste Politics* (Routledge); and *Ethics and Technology Assessment*; and co-edited volumes: *Governing Shale Gas* (Routledge) and *Engaging Environmental Justice*. His research in the field of environmental justice is funded by Nuclear Waste Services; Research England; The Economic and Social Research Council; Euratom; The Foreign, Commonwealth and Development Office; and the National Institute for Public Health Research.

Routledge Explorations in Energy Studies

Local Energy Communities
Emergence, Places, Organizations, Decision Tools
Edited by Gilles Debizet, Marta Pappalardo & Frédéric Wurtz

Building Resilient Energy Systems
Lessons from Japan
Jennifer F. Sklarew

Regulatory Support for Off-Grid Renewable Electricity
Edited by Ngozi Chinwa Ole, Eduardo G. Pereira, Peter Kayode Oniemola and Gustavo Kaercher Loureiro

Northern Indigenous Community-Led Disaster Management and Sustainable Energy
Ranjan Datta, Margot Hurlbert and William Marion

Energy Policy Design in the South-Eastern Mediterranean Basin
A Roadmap to Energy Efficiency
Bertug Ozarisoy and Hasim Altan

Living with Energy Poverty
Perspectives from the Global North and South
Edited by Paola Velasco-Herrejón, Breffní Lennon and Niall Dunphy

Intergenerational Democracy, Environmental Justice and the Case of Nuclear Waste
Lee Towers and Matthew Cotton

For more information about this series, please visit: www.routledge.com/Routledge-Explorations-in-Energy-Studies/book-series/REENS

Intergenerational Democracy, Environmental Justice and the Case of Nuclear Waste

Lee Towers and Matthew Cotton

Routledge
Taylor & Francis Group
LONDON AND NEW YORK

earthscan
from Routledge

First published 2025
by Routledge
4 Park Square, Milton Park, Abingdon, Oxon OX14 4RN

and by Routledge
605 Third Avenue, New York, NY 10158

Routledge is an imprint of the Taylor & Francis Group, an informa business

British Library Cataloguing-in-Publication Data
A catalogue record for this book is available from the British Library

ISBN: 978-1-032-72801-8 (hbk)
ISBN: 978-1-032-72800-1 (pbk)
ISBN: 978-1-003-42268-6 (ebk)

DOI: 10.4324/9781003422686

Typeset in Times New Roman
by Apex CoVantage, LLC

Contents

Introduction

In an age of climate emergency, nature loss, pollution, and resource depletion, questions of environmental justice become paramount. These questions regard the gaps between those that reap the benefits of production and consumption and those that bear the risks and costs and have become central to political decision-making – from global regimes to local planning. Environmental justice and the related concepts of climate and energy justice analyse and assess these processes. These justice frameworks explore how policy authorities, environmental specialists, and communities can best alleviate the imbalances of risk, cost, benefit, and decision-making control between beneficiaries of productive and consumptive practices and the marginalised communities living in the sacrifice zones of a global extractive economy.

Many of the environmental problems that face humanity today have been labelled wicked problems (Head 2022). These problems are distributed in cause and effect and over space and time, they are difficult to deal with as have the inertia of infrastructural imprints and the associated vested interests, and they involve deep uncertainty and ambiguity. This means that there are no simple, technofix solutions because these problems are distributed across borders and political cycles, and are as such hard to manage with the short-termism and state-centric nature of the current world system. These problems cause complex and serious dangers to human and more-than-human health and well-being. This book focuses on two such related wicked problems, namely the broad climate and ecological crises (CEC) and the issue of nuclear waste. Both issues evade simplistic technical solutions – not for want of both material and rhetorical endeavours – and evade nations' borders and jurisdictions. They are problems now, while presenting long and potentially damaging legacies into the future. Therefore, both these wicked problems are fundamental questions of justice. Of specific concern within this book are the twin, inter-related challenges of *intergenerational justice* concerning the balance of risks, costs, and benefits between members of different generations across time, and *intragenerational justice* concerning this balance across current populations and societies.

DOI: 10.4324/9781003422686-1

Although now a core concern across applied philosophy and the social sciences, environmental justice has its roots in the collective action of the US Civil Rights Movement. The Civil Rights Movement spurred protest, civil disobedience, and direct action campaigning to shed light on the unfair treatment of Black communities in the siting of hazardous waste facilities, pollutant controls, and risk-bearing industries in a way that compounded their political and socio-economic marginalisation within US civil society. Environmental justice connected the rights of Black communities for fair treatment under the law, to the burdens and costs generated by living and working under poor environmental conditions (Murdock 2020). Although initially situated within a particular US political geography, the scope of environmental justice action and scholarship later broadened to the treatment of marginalised people adjusting to the dangers of mounting climate and ecological crisis, not least indigenous communities suffering the injustices of settler colonialism.

Environmental justice as a political movement has spurred substantive academic analysis of the ways through which powerful actors can and should redress imbalances of risk and benefit distribution and ensure fairness and justice in the processes of decision-making. As Lockie (2018) notes, environmental justice scholarship must walk the fine line between evoking appropriate outrage and inspiring careful analysis. Philosophers and social scientists work in parallel to networks of activists, politicians, and civil society actors concerned with halting the global crisis in which the interacting limits of the Earth's biological and geophysical systems are exceeded, leading to declining welfare for all species on the planet. The publication of Our Common Future (Brundtland 1987), and the emergence of *Sustainable Development* as the hegemonic political discourse of environmental protection, introduces the notion that the needs of future people, communities, and societies must be incorporated into contemporary political debate about resource use, economic growth, and social development. As a point of practical philosophy, ensuring a fair future for the planet is a challenge not only of intragenerational justice between the powerful and marginalised but also of intergenerational justice between those alive today and those that will be alive in the future. Effective environmental governance requires us to examine the relations between present circumstances and the actions and capabilities of remote past and future generations. Questions of justice between these different societies across temporal horizons have become an important lens through which to examine collective industrial histories and toxic legacies, and for us to define and enact new technology programmes, policy institutions, and social practices of environmental protection. In this book, we explore mechanisms for elucidating intergenerational injustices and developing a long-term democratic framework for alleviating such injustices over long time frames.

Overall, we frame our analysis principally in relation to a broad array of liberal traditions within moral and political philosophy that have scaffolded debates on social justice since the 19th century. Liberal philosophical traditions of concern within this book include those stemming from Cartesian and logical positivist traditions that are discursively and materially reductive, emerging from a foundational binary of society/nature (Moore 2015) that has led to the subjugation of ecological systems. As Aldeia and Alves (2019) note, socioecological issues cannot be fully comprehended or resolved when viewed solely as an environmental crisis held in abstract from the social world. Instead, socioecological breakdown is itself a crisis of Western modernity and the types of worlds that can and cannot be created within its philosophical and political frameworks. Such liberal philosophical traditions include both rationalist and empiricist approaches – built upon foundationalist ontological positions that assert an objective world that can be classified, quantified, and described largely independent of the describer (Haraway 1988). Liberal philosophy also encompasses individualistic and cosmopolitan definitions of normative ethics, for instance human rights-based frameworks that discursively construct moral actors as individuals. However, this individualist construction is often independent from the historical and socioecological context in which individuals are embedded and ultimately depend upon. We posit therefore critique in response a range of liberal philosophical traditions that are broadly anthropocentric and individualistic, focusing upon human rights, measurable values, and utilitarian conceptions of societal wellbeing, while treating the ecological systems that support human life as common pool resources to use or sinks to dispose waste in (Winter 2022; Gelderloos 2022). In total, a combination of ontologically foundationalist philosophical traditions and individual rights-based approaches has led to material conditions that reproduce recursive ecological injustices over time, and it is this general philosophical approach that we critique within this book.

As part of this exploration, we will focus on three main areas that intergenerational justice discourse and the liberal philosophical approach defined previously have tended to neglect. These three areas also underpin a central argument of the book. This is that the systems, assumptions, and actors that create a problem are not the best suited to mitigate the problem. Or as Santos (2016, p. 19) said:

> The problem is that after five centuries of "teaching" the World, the Global North[1] seems to have lost the capacity to learn from the experiences of the World.

Firstly, we explore the role of children and young people in deliberation on the future. We focus upon the current and historic democratic exclusion of young people as non-political actors in democratic politics, and also their potential role as proxies for future generational interests. Children and young people present opportunities to voice new normative and practical insight into intergenerational justice problems and solutions. Secondly, we will focus upon the role of indigenous knowledge as a source of creative insights into thinking synergistically between justice now and into the future, and how alternative epistemologies can complement situated scientific knowledge on these key issues of environmental management. Grounded in a social epistemology of philosophical pragmatism, we posit scientific knowledge as one of multiple ways of conceptualising the future and the justice dimensions of that future. Finally, we will argue that treating *future generations* as a monolithic public devoid of social structure, cultural specificity, history, or agency is deeply problematic. This is because treating humanity as a monolith ignores ethnoclass, gender, ability, and other unjust divisions that characterise all nations to varying degrees today. Treating humanity as a monolith also implies a fixed notion of the human that is overdetermined now (Wynter 2003) but may also be overdetermined in the future, as this fixed notion of the human is projected backwards and forwards in time.

The first section of this introduction will define intergenerational justice and intergenerational democracy briefly. The second section will elaborate on the three themes above, giving brief examples and outlining the main questions and concerns that will run through this text. The third section will outline the book. The overall argument presented is that intragenerational and intergenerational justice do not necessarily compete – for a stable climate, and ecosystems with clean air, water, and healthy soils now will benefit future generations. Achieving this synergy requires a strengthening of *intergenerational democracy*, or involving children and young people, not only as legitimate political actors now who will inherit this world but also as proxies for future generations. Also, as a matter of epistemic justice, allowing different ways of knowing, from local and indigenous knowledge, allows us to present the scientific and technical projections of social futures as place-based and contextual – situated, needing local input and experience, and accepting of fundamental uncertainty. This, as the IPCC (2022) note, offers more just and transformative solutions. Finally, by focusing on and learning about the ethnoclass, gender, sexuality, and ability injustices of today and their roots in over 500 years of colonialism, we may avoid pushing these injustices into the future.

Defining Intergenerational Justice

The roots of a philosophy of intergenerational justice stem from the concepts of political obligations associated with the work of Edmund Burke. Burke broke from the classical liberal philosophy of Thomas Hobbes, John Locke, and Jean-Jacques Rousseau to define a society built upon a contract between individuals who exist in a "partnership not only between those who are living, but between those who are living, those who are dead, and those who are to be born" (Burke 1951, p. 165). Burke argued that political obligations and duties exist between those who belong to different generations, rather than simply those within the same generation. His work concerned the notion of an "Eternal Society," the ongoing link between successive societies defined by an iterative process of learning between generations – that human actions have long-term consequences and that it is reasonable not to prejudice the opportunities of our descendants by the actions that we take today (Gabor 2013). This conceptualisation of justice between generations as a form of political obligation is implicit in contemporary environmental policy-making. Decisions taken to limit greenhouse gas (GHG) emissions, to alleviate the burden of plastic pollution in the oceans, or to protect against nature loss are designed to ensure that our descendants inherit the world that is safe from climate emergency and rich in biodiversity and natural resources. The ethos of *stewardship* in which we are caretakers of the world, rather than its masters, underpins much contemporary action on environmental protection (Hart & Cornia 1994). Yet to the discipline of practical philosophy, understanding precisely what caretaking action we must take as a society (and who within that society bears responsibility for such action) remains a problem for political and economic actors.

The starting point for many definitions of intergenerational environmental justice is often the Brundtland Commission's statement on Sustainable Development: "Development that meets the needs of the present without compromising the ability of future generations to meet their own needs" (Brundtland 1987, p. 43). The understanding of future impacts as a relevant component of contemporary political decisions through a sustainable development lens is an expression of what Möckel (2023) refers to as a specific but historically contingent *chronopolitics* – where we balance between a fundamental *presentism* (see, e.g. Gumbrecht 2014) and *short-termism* implicit in our current world of conspicuous consumption, growth forecasting, and electoral cycles against socio-cultural practices of imagining and constructing the future, each laden with ethical questions (Adam & Groves 2007). Two of the ethical questions of concern in this book are: firstly, what do we mean by future generations? Secondly, what are these needs? The first question revolves around how generations are defined. We assert an overlapping generations model in this book, referring to specific birth cohorts who tend to have children about 30 years later and then eventually become grandparents – future generations here meaning unborn but also youth and children from the perspective of older generations. Fritsch (2018, p. 20) argues that this definition is important as it shows chains of concern and care that are "socio-ontologically more adequate and normatively more contentful." However, this definition also allows the differentiation of unborn generations from today's children and youth and seeing all alive today as in some way collectively responsible for the unborn. Fritsch (ibid.) adds two further senses we can use *generation* for. The second definition of generation involves the idea of shared cultural experience – for instance, generation *X* or *Z*, Millennials, or Baby Boomers. This definition of generation involves a shared sense of collective cultural norms, rules, and social circumstances, each handed down from previous generations via education, prevailing socio-economic conditions, and shared experiences of familiar cultural artefacts and historical events. Ultimately a fraction of this cohort becomes the economic, cultural, and political elite and thus takes a significant role in shaping society for the next generation. The third definition of generation is taken from a shared historical marker, notably the two world wars or other large global crises such as the Great Depression. Climate and ecological emergency (CEC) is the new and emergent historical marker that shapes this generation's lives and their outlook for the future. Overall, with this messy three-part definition, we have what might be seen as Janus-faced generations looking backwards and forwards and understanding both heritage and responsibility, able to learn from the past and shape the future in intelligent and caring ways (Watane 2022; Winter 2022).

The question of what we need now and in the future defies simply explanation and action. In the context of the CEC, we are demonstrably exceeding Planetary Boundaries, from greenhouse gas emissions in the atmosphere, biodiversity loss, pollution of the marine and terrestrial environment, invasive species, natural resource depletion, to growing crises over food and potable water availability (O'Neill et al. 2018). These socio-environmental challenges relate to things we need now and will continue to need in the future – a stable climate, fresh and clean water, fertile and healthy soils, and clean air. Oliveira (2022) deems these irreplaceable public goods and links their exploitation to contemporary economic decisions and material wealth, such as housing, and to more abstract social goods such as peace or justice. What all these things have in common is that they are common pool resources and sinks, owned by no-one and used by all, and thus can be degraded and diminished beyond repair or to tipping points that cannot be undone. Such public goods also cannot easily be substituted – the economic and material element of any environmental good relies upon extraction from and waste into the biosphere, so any use of these goods is subject to interconnecting limits. Thus, Oliveira (2022) argues we should take a Strong Sustainability approach that sees certain things and processes as irreplaceable and entailing a duty to maintain and nurture them irrespective of their short-term economic value to humans. This in turn implies a sufficientarian approach that guarantees a minimum quality of life involving these three interconnected spheres of social, material, and environmental goods. In this book, we argue that humanity requires a ceiling of resource consumption and a floor. This is because it is not those who are marginalised by class, economic status, ethnicity, culture, or indigenous community who have pushed society past its planetary boundaries to a point of global crisis. It is demonstrably evident that the dominant ethnoclass and the top 10% in income and wealth (Kenner 2019; Chancel & Piketty 2015) are the key actors responsible for driving planetary change. Yet the risks and burdens of our mounting ecological crises are borne by those that lack the economic and social capacity to adapt to the changes wrought by those in a position of economic and political power to do so.

Three Features of Intergenerational Justice

Children as Proxies for Future Generations

In contemporary democratic societies, children and young people have an ambiguous legal and political status. They are often treated as wards of their parents, carers, and the state, so certain human rights remain truncated or absent. Conversely, politicians hold young people as standard bearers of an imagined future culture, representing collective societal hope towards fixing the problems of the past. This inherent contradiction between the burden of future responsibility for the social good and the lack of decision-making control is a core concern of intergenerational justice within this book. The future direction in which we move, and the trajectory that shapes the world that young people will inherit, is categorically beyond their influence and input because they are legally not-quite-fully human. This contradiction is not beyond young people themselves. Collective social movements such as Fridays for Future (FfF) have begun to make demands specifically highlighting the injustice of current policy-makers' dreadful track record dealing with climate catastrophe, and the exclusion from decision-making of the voices most strongly affected by the crisis happening around us (Knapp & Renn 2022; Davies et al. 2016).

Theorists working with children's and young people's statements have argued that these people should be treated as proxies for future generations (Knapp & Renn 2022; van Dijk 2021a, 2021b; Spannring 2021; Davies et al. 2016). This serves to not only mitigate a democratic deficit now but also face the future and unborn future generations, implicitly tying together intra- and intergenerational justice. Davies (2012) developed the idea of *intergenerational democracy* to serve this end, aiming to include in substantive ways both young and older people who are routinely, if in different ways, diminished or discounted in modern "democratic" and less-than-democratic systems. This discounting and diminishing of young people is paralleled by injustices of ethnoclass, gender, sexuality, and ability (Fraser 2021), which results in attenuated, performative forms of democracy that exclude certain identities, ideas, and voices from political debate. These performative democracies seem unable to deal with wicked problems caused by the manifold socio-environmental contradictions of these societies. Thus, we must look beyond these performative democracies for solutions.

Indigenous Societies and the World System

From the end of the great empires of the 20th century, coloniality and post-colonial development continue to shape all aspects of socio-environmental change. Settler colonial societies instigate multiple and overlapping injustices and indignities upon indigenous communities, spanning from the routine denial of rights to deeper epistemic injustices denying alternative ways of being and knowing (Winter 2022; Paradies 2020; Estes 2019; Powell 2018). Also, those marked by the Global Colour Line (Lake & Reynolds 2008) and subjects/citizens of former colonies are still largely those who make up much of the global poor (Paradies 2020; Quijano 2010). We reframe poverty as impoverishment, as the majority world's poverty is a result of the minority world's actions and extractions (Koram 2022; Táíwò 2022; Hickel et al. 2022; Pettifor 2019). The CEC will exacerbate these injustices by making places in the majority world harder if not impossible to live in as the irreplaceable good of a safe and stable climate recedes. Despite this, the minority world's GHG emissions continue to rise, in effect colonising the atmosphere (including through excessive consumption-based emissions embedded in the luxury goods that they consume) and removing the space for economic development through energy use in the majority world (Hickel 2020; O'Neill et al. 2018). Furthermore, within societies in the minority world, especially the United States and United Kingdom, there is routine and systemic ethnic discrimination against racialised citizens and communities (Koram 2022; Táíwò 2022). This internal oppression is paralleled by a punitive and hostile border policy that targets, profiles, and diminishes the very people fleeing from the climate chaos caused by the minority world and mostly its dominant ethnoclass (Koram 2022; Walia 2021; Mbembé 2019). Thus, coloniality refers to these interacting injustices.

Despite more than 500 years of oppression, Indigenous knowledge and societies continue to offer positive visions for the future and potential solutions to broader problems embedded within the CEC. A foundational aspect of these visions and solutions is a relational ontology that sees humans as a (relatively minor) component within the wider web of life (Kimmerer 2013). This thinking overcomes the Cartesian binary of society/nature that so damagingly sets us apart from the world we rely on and are a part of (Moore 2015; Quijano 2010). A relational worldview, placing the humans interdependently within the web of life, allows a more humble and caring treatment of the animate and inanimate neighbours we share the world with. In terms of intergenerational justice, e.g. Watane (2022) describes how Māori societies face the past and their ancestors in a close and mutual recognition that they exist in a chain of human and more-than-human intergenerational relations, purposes, and responsibilities. Facing the past and one's ancestry leads us to face one another in the present, recognise one another, and collectively learn from past struggles, fears, and failures. This process leads us to face the future with renewed *moral imagination*, connection, and responsibility to recreate the world fit for future generations (Watane 2022). This perspective that sees synergies between intragenerational and intergenerational justice is shared and affirmed by many indigenous groups (Watane 2022; Engle et al. 2022; Estes 2019).

A further solution offered by indigenous societies is a potential renewal of our unresponsive and maladaptive representative democracies through direct or strong democratic means (e.g. Horn-Miller 2013). Whyte (2017), for example, describes how his Native American society practised a form of dynamic seasonal governance or a seasonal round, which appears to have been much more widespread before the imposed and rigid nation state model standard now (Sioui 1999; Graeber & Wengrow 2021). In this dynamic direct democratic model, leadership and authority take on very different aspects, with the former role of leader becoming one of responsibilities and duties and controlled by various countervailing mechanisms, while authority emerges for specific times and processes for specific and limited purposes, such as hunting, and then is replaced by more horizontal and non-hierarchical relations (Sioui 1999; Whyte 2017; Graeber & Wengrow 2021). A lesson here is that democratic decision-making can be done through very different, non-hierarchical, egalitarian, and more socially responsive ways than in the aggregative and representative democracies of the minority world, which have broadly regressed towards narrow neoliberal (or else populist) political worldviews in the late 20th and early 21st centuries (Koram 2022; Dean & Zamora 2021; Mbembé 2019). Historically, representative democracies were less a final stage in some teleological social development, more imperfect, and incremental compromises by elites endeavouring to limit popular power (Malm 2016a; Mitchell 2013). Finally, democracy is not some Western invention but a social good that has agonistically emerged time and time again in different forms, times, and places, often from below or from the periphery (Graeber & Wengrow 2021; Sen 2009; Graeber 2007). As such, democracy is something we should struggle to pass on to future generations. However, this democracy should not be top-down, static, and fixed or limited, but bottom-up, adaptable, and justice-oriented.

Humanity, Ethnoclass, Ability, Gender, and Sexuality

When we conceptualise human–nature relations within the Anthropocene, whereby humans have become the dominant influence upon environmental change, intergenerational justice discourse is largely blind to the manifold unjust divisions of gender, ethnoclass, ability, and sexuality that characterise all modern states to varying degrees. This is problematic for several ethical but also practical reasons. Most broadly, there is a risk that intragenerational justice and intergenerational justice become presented as a zero-sum game in which current and future well-being are deemed mutually exclusive endeavours. This would support an unjust status quo that without recognition risks extending these injustices into the future. More practically, an aspect of this status quo is that of *technocracy*, the placement of *expertise* in primacy in the resolution of socio-environmental challenges, and prioritisation of certain types of scientific, technical, and economic knowledge within that framework (Centeno 1993). Technocratic governance also channels policy solutions to environmental problems towards technofixes that, as will be shown later, either push problems into the future or spatially aside. As such, technocratic governance can be seen, as argued by Cotton (2017), as using a form *of synoptic rationality*. This form of rationality claims scientific and technical justification, but in reality it is a decision that is made prior to this authority with which it is subsequently adorned to add rhetorical force (ibid.). The practical problems with this form of governance are its limits and inability to deal with wicked problems like the CEC and nuclear waste, which arguably require just and democratic solutions.

More specifically, as argued earlier, the global top 10% in income and wealth cause 50% of GHG emissions (Oswald et al. 2020; Chancel & Piketty 2015). This outsized impact and extreme wealth of those in the 1% and above are often used to slow, prevent, and/or shape to their own purposes the solutions to the CEC (Dafermos et al. 2021; Aronoff 2021). The mere existence of billionaires in a world of finite resources, and scarcities imposed by neoliberal capitalism (Hickel 2020), is a powerful symbolic force that denies the potential for justice (Mcqauig & Brooks 2013). The technofix solutions on offer from the UK government, while rhetorically claiming to *leave no one behind* (Ofgem 2019), have not stemmed the tide of impoverishment and economic inequality in this country (Caddick et al. 2022; Marmot et al. 2020). Furthermore, treating the CEC as a technical problem leaves many people marked by ethnoclass, ability, gender, and sexuality difference lacking the funds to reduce their own GHG emissions. At the same time, these marginalised groups pay disproportionately more and/or do not receive the benefits more affluent people receive.

For example, electric vehicles (EVs) are offered as a technical solution to reduce emissions from transport. Governments around the world are offering subsidies to improve the market uptake of EVs and to encourage innovation in battery technology. This is a perfect example of a technofix *displacing* socio-environmental problems in space and time rather than resolving it. At the local level, car ownership levels of the most deprived are low, and these people tend to live in the most air-polluted and road accident-prone areas (Lucas et al. 2019). Accidents and pollution seem at least partially independent of the type of vehicle driven. Accidents can occur in EVs not least due to their quiet nature, and much of the pollution in the PM2.5 scale likely to be damaging to health comes from tyre friction and brake wear (rather than tail pipe emissions) (Carrington 2022; Beddows & Harrison 2021). The actual emission profile of an EV is approximately twice as much embedded in the build than that of fossil fuel vehicles, and its overall sustainability is dependent upon the prevailing energy mix at the national scale of electricity supply (Cerdas et al. 2018). EVs present an individualised and consumerist solution to a collective socio-environmental problem – low-carbon mobilities – that can be better managed through changes to planning, public transport, walkability, and cyclability of urban and peri-urban environments (Marx 2022). The mining processes for the copper, rare earth minerals, and lithium essential to the EV batteries involve injustices, such as children mining in extremely unsafe conditions, and form part of a minority world and majority world unjust flow of wealth and illth. Sovacool et al. (2020) call these flows the decarbonisation divide – presenting a fundamental challenge to the achievement of a global *just transition* that simultaneously secures worker rights and well-being while shifting towards sustainable modes of production and consumption (Newell & Mulvaney 2013). Temporally, these flows of low-carbon wealth and illth are set to soar by orders of magnitude in models that include the widespread role out of EVs (Sovacool et al. 2020). This injustice of the decarbonisation divide is thus a continuation of a longer history of neocolonialism and extractivism that endanger the most vulnerable workers across the developing world (Gelderloos 2022; Táíwò 2022; Dunlap & Arce 2021).

In this brief example, we see how techno-fix solutions ignore and occlude myriad injustices. Technocratic governance also relies upon *synoptic rationality* – seeking not the optimal solution to resolving future harm but rather a continuity of individualised solutions to a fundamentally collective action problem. Finally, technocracy pushes injustice onwards, horizontally in space and vertically in time. As we argue in this book, denying justice now leaves little hope of building solidarity and care for seemingly abstract future generations. Rather we should explore how working toward socio-environmental justice now can create long-term solutions and synergies between inter- and intragenerational justice.

Book Outline

The first part of the book covers what we call the old liberal ontology and praxis that have created the CEC and seems only able to load the costs of these crises onto its many others. This part of the book also explores possible new ways of thinking, acting, and being that may get us out of these crises and offer more substantive and inclusive forms of justice. The first chapter explores the range of philosophical approaches to intergenerational justice. It surveys the intergenerational justice ideas and problems raised by liberal philosophy, including the non-identity problem, issues around the idea of reciprocity, and social discounting. It then goes on to cover how the more progressive stands of liberal philosophy including sufficientarianism and environmental rights go some way to tying intragenerational and intergenerational justice together. However, these approaches are still limited by liberalism philosophy's exclusions. The chapter then goes back to underline the major of the pathologies of liberal philosophy that make it difficult to deal with justice substantively, including its Cartesian focus on forms/objects over flows and relationality, its individualist fetishisation, and its anthropocentric nature.

Chapter two then follows minority figures, from Spinoza who challenged Descartes with his pantheism, through Marcel Mauss and later anthropologists' considerations of the social relations of the gift. The chapter then considers indigenous philosophies. These philosophies appear to understand the problems with liberal philosophy as these people have lived as the Others under liberalism. As such they've suffered under its pathologies by being placed on the wrong side of the many Cartesian divides. Finally, the chapter ends with a consideration of the human and how we have and could see this over deep time.

Chapter three explores the empirical emanation of liberal philosophy, namely mainstream economics and its treatment of intergenerational justice and the CEC more generally. Regarding intergenerational justice, economists generally see this as a trade-off with intragenerational justice and thus devise methods of pushing the problems and costs onto the future generations under various justifications. The main justification being continued economic growth means all things will be better/richer in the future, so we can leave decisions and actions to the future – a highly dangerous gamble to say the least. This leads us to a consideration of how mainstream economics deals with the CEC and the epistemology and methods behind this. As various theorists have shown, mainstream economics has narrowed its epistemology and attendant methods even as it gained greater and greater influence. The results are dangerous ideas produced from highly questionable methods, suggesting mainstream economics should not be influencing policy surrounding the CEC, or arguably any other public policy.

Chapter four moves onto newer ideas and approaches surrounding intergenerational justice. The first approach is a reworking of game theory prisoners' dilemma as an intergenerational dilemma that includes some interesting techniques for elicitation of intergenerational empathy. The second group of approaches examined explore the framing of children and young people as proxies for future generations. This section draws these approaches together through the method and methodology of intergenerational democracy. This method of intergenerational democracy is in turn connected to an adapted form of the capability approach in order to outline a more substantive and inclusive form of justice. Finally, the chapter considers the problems with future forecasting and scenario thinking that work with ideas of *probable*, *feasible*, and *realistic* over backcasting futures that are *desirable*, *just*, and *liberational*. We argue that this form of decolonised backcasting would be more suited to our substantive and inclusive form of justice and a tool or method that can be used to move towards this goal.

The second part of the book moves onto the cases of nuclear waste. The nuclear, via its spatial and temporal reach and impact, is a quintessential issue of inter- and intragenerational in/justice. To understand this reach and impact, Chapter five lays out some theoretical tools used to analyse the nuclear, including the ideas of nuclear landscapes and communities, peripheralisation, energopower, and nuclear colonialism. It is with these theories and critical approaches that we assess and analyse the following three case studies.

Chapter six focuses on Nuclear Waste Management (NWM) in Canada, which on the surface seems a good example of deliberative processes, including the use of indigenous knowledge and participation. However, when looked at more carefully, the historical wrongs have not been dealt with justly in any way and can be seen as a case of nuclear colonialism. Also, the present search for a geological disposal facility (GDF) involves issues of elite capture (a process whereby local elites intentionally or unintentionally dominate a local process, occluding other voices and interests). Thus, the Canadian Nuclear Waste Management Organisation (NWMO) does not live up to its own stated ideals and has elements of peripheralisation.

Chapter seven examines the case of Finland, the state closest to achieving its nuclear waste solution. This, on the surface and according to the Finnish nuclear authority's rhetoric, seems to be a progressive and just affair. However, again when we examine the historical development of this case, there are serious democratic deficits and issues of peripheralisation. This leads to what has been called divided communities accepting the GDF with resignation – one of the cultural aspects of peripheralisation and nuclear communities.

Chapter eight considers the case of the United Kingdom. Looking at the history of the United Kingdom's nuclear policy development, we see four periods and associated discourses, namely an early techno-optimist and securitisation era, a nuclear expansion and emerging socio-environmental problems era, a deliberative turn, followed by an era of climate change securitisation. Nuclear waste policy during this period followed suit with early top-down governance failure, leading to a much more analytic-deliberative approach requiring community opt-in to the GDF siting, which unlike nuclear power policy has retained its deliberative elements. We show this through an analysis of one specific Community Partnership and its published meeting notes. However, broader questions remain over who controls and shapes the actual democratic decision of this voluntary siting process. Also, a recent and significant expansion of government plans for new nuclear builds, which will entail a higher capacity inventory of the GDF, will likely place pressure on this deliberative process. We then analyse the costs of the GDF, nuclear decommissioning, and nuclear subsidies more broadly, finding these costs to be of significant and questionable social value. Considering these costs and the broader risks of the nuclear, in the final section we argue for a new democratic process to decide on our energy future.

The book concludes, firstly, with an overview of the environmental justice aspects of distribution, process, recognition, and reparation. Globally and nationally we are at a juncture, and justice demands a radical change in direction. This begins, continues, and ends with the radical redistribution of wealth and power. Politicians who do not recognise this must be dismissed as the anachronisms that they are. Liberal politicians and their ideas have failed and are failing us all, and it is time to move on. Our democracies need to be revived, and this requires new thinking, new processes, procedures, and new actors, especially the young who must have a say in their own future.

The second part of the conclusion takes a more epistemological approach, considering scientific knowledge and what it is good for and not so good for. Scientific knowledge is a way of understanding *and acting* on the world, and thus a process of discovery and creation. It is a very important way of acting on and understanding the world, but there are other ways that may complement science. Moreover, science can tell us what we *can* do but not what we *should* do. Thus, we propose science must become more systemic and relational, as it is in many fields we reference such as strands of ecology and biology. It also needs to widen its peer-community and democratise its processes, both in what it investigates and how.

Note

1 Throughout we use the terms minority world as Global North and majority world as Global South. These terms describe the fact of who and how many live where, while also pointing towards the injustice of minority rule over the majority.

Part One

Intergenerational Justice Dilemmas

1 The Philosophical Challenge of Intergenerational Justice

This chapter will outline the main thinking and philosophical approaches to intergenerational justice and the limits of this thinking. A series of problems has been identified in this literature that basically revolves around the one-way direction of communication and information over time. More specifically, this chapter looks into the non-identity problem, the problem of reciprocity, and social discounting. Then as answer to these problems, it examines the theories of sufficientarianism, the capability approach, and environmental rights. Taken together, these approaches offer solutions to the problems of intergenerational justice, but only within the limits of an individualised and anthropocentric liberal philosophy. Going beyond these limits for more substantive and inclusive forms of justice will be the subject of the next chapter.

Philosophical Challenges and Concepts in Intergenerational Justice

Intergenerational justice has two temporal components. The first typically refers to relations between different age groups: the relations of people whose lifetimes overlap but for whom the different stages of the life course present different issues of power, economic opportunity, welfare, and identity. The second typically refers to intergenerational relations, in which future and past generations can be seen as holding legitimate claims or rights against present generations, with concomitant duties to future or past generations, but where such lifetimes do not overlap. Under the second definition, the current generation refers to all people currently alive irrespective of their individual ages. In this book, we primarily consider the second aspect of intergenerational justice. However, when it comes to questions of democratic accountability and control over contemporary decision-making that affects future lives, these two aspects of justice remain intertwined.

As a philosophical problem, intergenerational justice concerns three principal issues. The first is a meta-ethical question over whether distant future people can be the bearer of rights expressed towards current people. The second relates to what obligations current generations can and should hold

DOI: 10.4324/9781003422686-3

towards both past and future generations and what weight we give to fulfilling obligations towards them. The third concerns the practical steps that would need to be taken to alleviate injustices across time as they are identified by current generations (Meyer 2017). We take each of these issues in turn in the following sections.

Can Future People Have Rights? The Non-identity Problem

The first meta-ethical concern is sometimes described as the *non-identity* problem identified by Parfit (2017). The non-identity problem concerns how action in the present affects future generations, raising doubts over the possibility that currently living people can substantively harm future people's interests.

The starting point is that of harm. When person *A* harms person *B*, we can work to show how person *B* is worse off than if person *A* had not carried out such an action (or had chosen a different course of action in which no harm occurred). This harm has a temporal dimension: it can be measured from the point of action across a timescale in which both individuals *A* and *B* are identifiable. This harm may have a short-term or long-term impact upon person *B*: in any case, we know that person *A* has violated an obligation to ensure that harm does not befall person B because person *B* has an inalienable right to live free from harm. We also know that for justice to be observed in the event that person *A* is culpable for harm to person *B*, an institutional framework must exist for redress in which person *B* provides compensation or justice through some other means. Such thinking is common to an understanding of Berlin's (1957) concept of *negative liberty*, in which freedom extends only so far as it does not hinder the freedom or welfare of others, and thus of reciprocity, in which obligations are shared amongst all within society to ensure that the rights of others are respected.

The non-identity problem is an expression of the difficulties experienced in measuring harm across a temporal horizon in which person *A* and person *B* never meet. Actions taken by person *A* may have consequences that harm person *B*; however, because at the time when the action was taken there *was no* person *B*, no harm could be said to have occurred. Any action taken today not only changes future living conditions for future people but also simultaneously changes who those future people *are*. Both the number of people alive in the future and the collective differences that would emerge in genetic make-up, upbringing, identity, and community are all shaped by current actions. There is no fixed set of individuals waiting to be born, but rather the composition of a future society is contingent upon the actions of those alive today. If person *A* carries out a harmful action, that means person *B* does not then exist in the future, then person *B* cannot be said to be harmed by such an action. Such a problem is similar to that described as the "grandfather paradox" (often used as a plot device in time travel fiction) – that if the past is changed in any way, this creates a contradiction: if one travels to the past and intervenes in the conception of one's ancestors, for example, by causing the death of a grandparent before they can conceive the protagonist's parent. The time traveller would not be born and thus could not undertake the action in the first place. One can apply similar thinking to the rights of future people – any action, whether harmful or not, will change the very existence of future individuals. This non-identity creates problems for environmental decision-making.

Of note is Parfit's (2017) discussion of the issue of climate change and future generational interests. Parfit considers climate change policy under two scenarios. Firstly, society has the choice of two mitigation policies – one more expensive but rapidly reducing greenhouse gas (GHG) emissions, and the other is business as usual – increasing short-term economic growth but not significantly reducing emissions. If society collectively choses the initially cheaper, business-as-usual policy, the resulting climate chaos will increase risks to future people – from droughts, floods, hurricanes, heat waves, saline intrusion, biodiversity loss, and so on. Nevertheless, one cannot say that the future generations alive in the society produced by the first strategy would be better off than those in the society produced by the second strategy as the specific future people of the first timeline would not exist or be the same as the future people in the second timeline (Parfit 2017). The two futures are incommensurable. Thus, the crux of this problem is that one can never compare two or more future policies, as simply enacting one policy changes the future and the people who would be in this future. Or, the creation of climate chaos, as we are currently doing, *actually constitutes particular future generations* and their very existence, and so they will be in no position to complain (Fritsch 2018). At the extreme end of this argument, in an era of environmental emergency, some are concerned that current actions will lead to *no* future people – an extinction event for the human species. Although this would undoubtedly be a morally catastrophic outcome, the future people cannot be said to be harmed by this action as one cannot hold rights if one does not exist (Beckerman & Pasek 2001).

What Obligations Do We Hold to Future Generations? The Problem of Reciprocity

Reciprocity is a principle of exchange. Such exchange is primarily one to establish cooperation and mutual benefit. Reciprocity is a guiding principle for social interaction. The most basic form is *direct reciprocity*, whereby individuals exchange benefits directly with one another. If person A helps person B, this then creates a social obligation for person B to help person A in return when needed in the future. As Social Exchange Theorists such as Homans (1974) and Blau (1964) argued, social interactions and relationships can be expressed in terms of costs and benefits exchanged between individuals or groups – whereby people engage in social interactions with the expectation of receiving rewards and avoiding punishments proportional to their relative level of investment. This tit-for-tat reciprocal agreement is the foundation of contractual obligations amongst peers – the exchange of mutual benefit only works if both parties have transparent agreement on the terms of the interaction. *Indirect reciprocity* is a more complex extension of this phenomenon, in that individuals observe and reciprocate based not upon personal need or circumstances but rather through witnessing the reputation or behaviour of others. For example, if person A witnesses person B helping person C, person A may be more inclined to help person C in the future after having observed the mutually beneficial exchange. *Indirect* or *reputation-based* reciprocity is about the establishment of social norms of helping, giving, and exchange when personal gain is not a direct consequence. Both direct and indirect reciprocity are used in classical economics (including game theory) and behavioural economics to explain how cooperative behaviours and other relevant social norms around fairness and mutual benefit explain trust in economic transactions (Falk & Fischbacher 2006).

The extension of indirect reciprocity to society at large towards a more common understanding of social justice is sometimes referred to as *generalised reciprocity*. Generalised reciprocity is the principle of giving without the expectation of something in return from any specific individual or group. Generalised reciprocity is the foundation principle of broader social trust – that establishing a norm of giving, charitable action, or self-sacrifice to non-identified individuals contributes towards general societal well-being even if the individual does not benefit directly as a result of their own actions and investment. This generalised reciprocity is vital to maintaining community relations, building interpersonal trust, and enabling cooperation. It links social norms of fairness, altruism, and generosity. It is against this backdrop that contemporary thinking around intergenerational justice is established. The concept of *futurity*, through which an individual contributes to general well-being of the future irrespective of any personal gain, is an expression of generalised reciprocity over time (Groves 2009). Reciprocity is thus relevant to a broader social contract theory concerning fairness within society, relating both to human rights and to the distribution of welfare benefits, risks, and costs. Rawls (2009), of note, discussed distributive justice through a *difference principle* – that inequalities should only be permitted in a free democratic society if they benefit the least privileged more than an equal distribution would. Rawls described a principle of generalised reciprocity as an intergenerational concept through a *just savings principle*. This principle focuses passing on just institutions, excluding the distribution of wealth beyond that needed to ensure the continuation of just institutions, with the further condition that this just savings principle would be mutually agreeable to any one generation (Rawls 2009, 1993). Each generation is obligated under the just savings principle to save a fair share to preserve just institutions into the future on the assumption that all generations saved in accordance with the same principle, each collectively agreeing to leave a normatively "better" society than the one that they were born into – a Golden Rule between generations "Do unto others as you would have them do unto you," irrespective of individuals within each generation (Gabor 2013).

Reciprocity links to the non-identity problem, in that a narrow or direct concept of reciprocity makes intergenerational justice impossible. Direct reciprocity-based justice would require "that persons provide benefit for others, including members of different nations or generations, only if the recipients are in a position to reciprocate" (Page 2007a). If there is no social contract between current and future generations due to a lack of specific identity and shared mutual cooperation between different individuals, then non-reciprocity becomes a barrier to intergenerational justice. However, a more generalised concept of reciprocity, in which actions are motivated by total conception of future well-being irrespective of individual investments and outcomes, is a stronger starting point. McCormick (2009) argues this generalised reciprocity allows us to pay attention to our *intuition* that current actions that threaten the well-being of future people violate the requirements of justice, even in the face of the non-identity problem.

The Weighting of Future Obligations: The Issue of Social Discounting

Rawls' concept of a just savings principle requires a mechanism for weighing our social obligations between current and future generations. Questions arise over the appropriate calculus between intragenerational and intergenerational justice components of contemporary policy decisions. In essence, we must find a suitable balance that reflects society's preference for consumption timing, balancing the weight we collectively place on present versus future consumption. Rawls' overall theory of justice weighs intragenerational relations above intergenerational relations, justifying this on the grounds of there being natural reciprocity between people in the present (2009). This theoretical underweighing of intergenerational justice has had a performative effect on economic and natural resource policies, with many states applying a discount rate on future investment. In economic terms, a discounting function is applied, whereby we reduce the value of future costs or benefits to their present value. Discounting is based upon the principle that people generally prefer to receive benefits sooner rather than later, and costs are typically seen as less burdensome when they are in the future. The process of discounting involves applying a discount rate to future costs and benefits to calculate their present value. The effect of this is the more distant in time, the less monetary worth is placed on this investment, or this investment depreciates according to the discount rate.

For political decisions (rather than private financial transactions), policy-makers commonly apply some form of a *social discount rate*, that is, the discount rate used to calculate the present value of future costs and benefits from the perspective of the society as a whole (Caplin & Leahy 2004). Unlike private discount rates which individual savers or businesses might use to make investment decisions, the social discount rate considers broader societal welfare and intergenerational equity in the calculus. Determining the appropriate social discount rate for future generations is ethically contentious involving many non-economic normative considerations. Different stakeholder groups may have varying views on how much weight should be given to the well-being of future generations compared to the present generation. Some may argue for a low discount rate, reflecting a strong concern for the welfare of future generations, while others may advocate for a higher discount rate, prioritising present consumption and economic growth to meet immediate needs. As Caney (2014) argues, we must clearly specify whether we accord equal moral status to all persons irrespective of their temporal position (i.e. in the present or future) and then question if it is appropriate to ascribe a lower weight to human interests the further they are in the future based solely upon the fact that they exist further in the future. This is described as a *pure time preference* within the discounting of future lives against current lives, such that a preference is expressed for something to come at one specific point in time rather than another merely because of when it occurs in time (Lowry & Peterson 2011). Such pure time preference is based upon a series of assumptions: generations naturally connect and overlap, so there is no need to think about future generational interests; a concern for future people is unjustified as they will naturally be better off as they benefit from accumulated knowledge and technology; intergenerational and intragenerational justice are trade-offs, and thus benefiting future people over disadvantaged present generations is unjustified; and the inherent uncertainties of what might happen or not happen due to ongoing CEC, and our collective ignorance of what future people will value or want means it is pointless to try to consider distant generations' interests (Thompson 2010).

At the core of social discounting, reciprocity, and non-identity considerations are a set of overlapping meta-ethical questions. Firstly, are the economic principles used to establish the value of natural resources, financial capital, and investment decision-making ethically appropriate when applied to future lives and future well-being? Secondly, can we establish a right-based approach to future generations if they have no reciprocity with the present? Thirdly, does the existence or non-existence of future people abrogate our responsibility towards ensuring long-term environmental stability, resource availability, and institutional good governance into the future? Shue (2014, p. 230) suggests that while the above considerations ponder important considerations such as the size of or specific characteristics, values, and needs of future generations, this does not "much affect the general shape and character of our responsibilities to whichever human beings turn out in fact to live in the future." This is because if one has any responsibility to humans whose lives one can significantly affect, then one has responsibility to any future humans regardless of numbers or identities (Shue 2014). This general point that resolves the problems of non-identity, reciprocity, and discounting can be supported more fully by consideration of objections and solutions to each of these three problems. However, it is important to note here our core argument that these philosophical problems all tend to stem from an overall liberal individualism that militates against social solidarity, whether that be intragenerational or intergenerational.

The non-identity problem has been described as the "shadow out of which all work on future duties struggles to emerge" (Fritsch 2018, p. 33). Much of this struggle emerges from the concept of harm and how this is then attached to specific people or groups, in what might be seen as in the social contract liberal tradition. The problem is that those future generations that will exist would not exist if the harmful action of present generations was not taken. It is of note that Parfit (2017) himself argues this non-identity should not matter, but rather it was a lack of ethical theory to cover this issue. Two connected approaches are useful here – the sufficientarian and the environmental rights approaches, which along with arguably resolving the non-identity problem also sidestep the problems of non-existence, non-reciprocity, and future discounting.

Sufficientarianism, or Is Enough, Enough?

Sufficientarianism is a branch of moral philosophy that expresses justice in terms of basic minimums of human life, health, fulfilment, and dignity. Brock (2018) argues that everything concerning contemporary discussions of distributive justice is deeply contested. Understanding a fair distribution of benefits and burdens will need to specify which benefits and burdens are relevant, what account of fairness is being assumed, who is owed justice, and what the scope of that justice ought to be (ibid., p. 86). *Egalitarian* accounts of justice are concerned with equality of relevant benefits where individuals should end up with roughly the same level of resources, well-being, or opportunities. Egalitarian principles aim to reduce or eliminate disparities in material conditions and social advantages across society – either through the redistribution of resources from more advantaged to less advantaged members within a society or by ensuring that all individuals have an equal chance to succeed and pursue their goals. The former is a redistributive egalitarianism, and the latter is concerned with equality of opportunity. Within egalitarian philosophy are *prioritarianist* approaches that assert that it is more important to provide benefits to those who have less. In contrast, *sufficientarianism* focuses upon the claim that justice requires that everyone get an adequate or sufficient amount of specific goods and that aiming at equality beyond that sufficiency threshold is either a second-order priority or an inappropriate focus of justice. Frankfurt (1987, p. 23) wrote: "If everyone had enough it would be of no moral consequence whether some had more than others." Any political or economic decision will affect the distribution of benefits and burdens within a society. Some will be better off and others worse off after any decision is made. In Frankfurt's concept of sufficientarianism, he asks us to assess the possible decisions with respect to whether or not people would have *enough* in the distribution as a result, rather than focusing solely upon who gets what and the equality of outcome from each choice. To proponents of this philosophical position, *sufficiency* is not simply instrumental (such as having enough income to allow an individual to eat, have somewhere safe to live, and to heat that home); rather, it represents a level of respect for the agency of individuals – that sufficiency is a morally fundamental threshold (Shields 2020), although, of course, defining what that threshold should be is a key meta-ethical consideration.

Broadly, sufficientarianism has been proposed as an alternative to both strict egalitarian and strict utilitarian views. The strict egalitarian holds that all individuals should have equal access to resources, opportunities, and outcomes (i.e. advocates for the complete elimination of any form of inequality, whether it be in terms of wealth, income, social status, or opportunities). In contrast, the strict utilitarian view might otherwise overlook the importance of meeting the basic needs for human dignity if overall welfare is satisfied for the majority of people. Sufficientarianism as a philosophical approach has been adapted by capability thinkers in defining what level of individual needs forms the basis for human flourishing. A sufficientarian capability theory would allow for some inequalities beyond the minimum threshold as long as they contribute to lifting the least advantaged individuals above a sufficiency level. Page (2007b) argues that issues with egalitarian and prioritarian approaches can be overcome by the sufficientarian approach. Parfit (2000) distinguishes egalitarian approaches into telic – that argues any inequality is wrong – and deontic – only the inequality that has been caused reprehensibly is wrong. Page (2007b) claims this results in the problem of telic levelling down, whereby rich and poor groups are levelled down to no-one's overall benefit, while the deontic approach leads to leaving those suffering inequality, that is blameless, to fend for themselves. The prioritarian approach is non-comparative and focuses solely on mitigating the absolute disadvantage of the worst off. However, like egalitarian approaches, Page (2007b) argues this results in a fetishisation – here on absolute disadvantage in contrast to egalitarianism fetishisation of comparative disadvantage, with the result that both approaches lead to envy and pity and more general meddling in people's lives. Page (2007b) gives an example of where there are two social groups: one "very happy" and the other "extremely happy" – both egalitarian and prioritarian approaches would, for different reasons, intervene when this seems absurd from a sufficientarian approach.

Page (2007c, 2007b) notably draws out a sufficientarian ethic for global justice across time. He begins by pointing out such an ethical theory has to wrestle with three questions of justice: the *scope* or which entities are we considering – humans or humans and more-than-human, for instance; the *currency* or "the what" that is being saved/distributed be it resources or social institutions; and the *pattern* or the overall distributive aim be it equality, priority, or sufficiency. Page (2007c) explores *currency* of justice examining different approaches but settling on the capability framework of Sen (2009) and Nussbaum (2004). This framework has many advantages over rights frameworks in that they ask what people can actually do, and be rather than have the right to, which in many cases is an empty promise as there is a lack of opportunity to achieve or access this right (Sen 2009). The capability approach shifts the moral focus away from assessing simply material possessions or income, as is the case in traditional welfare economics, towards evaluating the freedoms and opportunities that individuals have to flourish, that is to lead lives that they value. The capability approach focuses upon the structural and relational factors that constitute what people can do and be – and function – the actual realisation of human capacities (Smith & Seward 2009). Rather than measuring this functioning in terms of resources or utility, a capability approach would emphasise the real opportunities and freedoms that individuals have at their disposal, recognising the importance of agency and the capacity to make informed choices about their lives, while respecting diversity in preferences and values that might lead to different choices or different outcomes. Moreover, the capability approach considers the structural social, political, cultural, and economic factors that might enable or constrain individuals' capabilities. Factors might include access to education, healthcare, facilities, clean energy, universal suffrage and political participation, and freedom from social discrimination.

Capability thinking places human well-being at the centre of environmental and social justice analysis. Nussbaum (2004) draws out a provisional list of basic capabilities that begin with life and end with political and material control over one's environment. Page (2007c, p. 445) argues that from this position we can focus on a healthy and hospitable environment not because it satisfies surface-level desires but because it is "an integral feature of a life of decent quality." Capabilities thinking thus recognises that environmental quality and access to natural resources are essential to all other aspects of human flourishing. By allowing access to clean air and water, the enjoyment of nature and cultural landscapes, clean energy, and pollution-free living spaces, these elements, in turn, reduce social, biological, and economic stress on the individual, allowing them better opportunity to take part in other aspects of managing their own well-being (such as by attending school or participating in political life). The intersectionality of the various dimensions of disadvantage, including ethnoclass, ability, gender, and geography, is equally important to understanding capability within the natural environment because inequality hinders access to good-quality environments. Marginalised communities will face multiple forms of oppression and disadvantage simultaneously, which in turn exacerbates environmental injustice – marginalised communities are disproportionately burdened by pollution and environmental hazards primarily due to the historical legacy of discriminatory policies and practices that trampled upon their environmental rights. The capability approach thus prioritises policy interventions that counter the path dependency of contemporary environmental injustices, by reducing inequality, addressing systemic discrimination, institutional racism, impoverishment, and promoting more equitable distribution of environmental resources and risks. Establishing environmental justice through a capabilities approach requires us to think of how an individual interacts with their environment; how environmental quality shapes political and social economic participation in society; and the dialectical relationship between these components.

One of the key challenges for applying a capability approach to intergenerational justice remains the non-identity problem. Assessing the capabilities and capacity of future people is necessarily participative. To understand an individual's needs requires us to ask the right questions about what they can and can't achieve within their current situation, and imaginative engagement with how best to overcome structural constraints in order to achieve sufficient agency for the individual. Assessing capability is, of course, subject to temporal uncertainty – predicting what capabilities will be valued or what challenges people of the future may face requires making assumptions about predicted social, economic, and environmental conditions. The longer the time horizon, the more difficult it is to consider the consequences of current actions and thus how present action shapes future capability. This should not be used as an excuse not to act. Indeed, a sufficientarian approach offers solutions. Sufficientarianism as an intergenerational theory presents a different *pattern* of justice. If the primary goal of a sufficientarian model of justice is to ensure that everyone has enough resources to live a decent life, prioritising basic access to necessities such as food, shelter, healthcare, social networks, education, and natural resources is a relatively simple future-focused policy agenda. By focusing upon the minimum thresholds through which no individual should fall, sufficiency is defined as the level of resources necessary for *any* individual to live with dignity and participate fully in society. This represents a minimum level of resources needed for an individual to be capable within society. The advantage of this approach is that it is easy to imagine – we can assess current thresholds of minimum environmental standards to meet current capability needs and then project this into the future. This means that inequalities in capabilities may be tolerated over an intergenerational timeframe if we can still ensure a minimum level of benefits for what we can predict will be worse off in the future. Nielsen and Axelsen (2017), for example, apply the sufficientarian approach to capability thinking by splitting capabilities into three groups: those related to biological and physical needs, those related to the fundamental interests of an individual, and those related to the fundamental interests of a social entity. Within each category, achieving sufficiency necessitates different distributional patterns amongst these different capabilities. These patterns vary based on how the capabilities operate and interact with one another, including over time. Thus, inequality between fundamental interests is acceptable as long as the overall pattern allows for individuals to meet the needs of their own capabilities. We can see therefore that arguing for a prioritarian or strict egalitarian approach that ensures equality between current and future generations in terms of environmental quality is deemed unnecessary, if we can ensure that current decisions will meet a minimum threshold with which future people can meet their needs for a dignified and autonomous life.

Oliveira (2022) develops the idea of a sufficientarian ethic by combining this with the ideas of planetary boundaries and the strong sustainability implication that certain social, environmental, and economic goods are irreplaceable. The first point is that we live in a finite world with interlinking ecosystem cycles that can be and are being pushed into crisis modes or tipping points from which they may not recover easily or at all (O'Neill et al. 2018). Thus, ideas of egalitarianism would be problematic in a near future, which likely requires the rapid reduction in overall levels of consumption (including natural resources and energy use). This returns us to the question of the currency of justice. Adherents to a weak sustainability philosophy argue that the solution to environmental problems is possible through substitution between different types of capital (natural, human, social, and manufactured) – that it is possible to maintain economic growth and reduce overall environmental impact by shifting from natural towards human capital as long as overall capital remains constant. Oliveira (2022) builds upon the arguments of strong sustainability theorists in recognising that some degree of *substitution* may be possible and necessary, particularly for human-made and social capital, but argues that there are limits to substitution when it comes to natural capital. Strong sustainability maintains that certain environmental and ecological aspects, such as biodiversity, could never be adequately substituted. These irreplaceable goods are divided into social (peace and democracy), environmental (clean water and stable ecosystems), and material/economic (commodities). As such, these goods are not only essential to a "minimum quality of life for any generation" but also susceptible to human overuse with dire consequences of their absence or depletion beyond certain levels (Oliveira 2022, pp. 8–9). In a sense, Oliveira (2022), by proposing the use of planetary boundaries and irreplaceable goods, implies a ceiling as well as a floor of human dignity and fulfilment but does not state this explicitly. This book advocates both floor and ceiling. If we accept that there are planetary boundaries that will threaten human and more-than-human life and livelihoods if certain thresholds and tipping points are crossed, then surely it is the excessive consumption and associated emissions that must be targeted. This excess seems largely located, reinforced, and protected, in the lifestyles, fossil fuel investments, and political power of the polluter elite (Kenner 2019), alongside more generally the extremely wealthy (Oxfam 2022; Oswald et al. 2020), the fundamental arbiters of global fossil capitalism.

Environmental Rights

To establish a sufficient threshold or floor of human dignity, environmental quality, and biophysical well-being, the concept of environmental rights is useful in protecting and ensuring that these needs are met in the future. Environmental rights are described by Hiskes (2005) in a way that parallels and complements the insights of the sufficientarian approach. Environmental rights contrast to human rights in that they are more broadly communitarian than individually focused. Hiskes (ibid.) argues that people have inherent entitlements to a healthy environment and to participate in decisions that affect the natural world. Such rights are inherent to individuals stemming from the fundamental value of human dignity and well-being. However, Hiskes (ibid.) also emphasises the role of interconnectedness between human rights and environmental quality. She argues that when environmental degradation occurs, this undermines an individual's enjoyment of other human rights, including the right to a decent standard of living and the means to meet their own capabilities and needs. Recognising the *intrinsic* value of nature and how that connects to the *instrumental* use of environmental resources by humans to meet their own capabilities is a core aspect of environmental rights thinking. One cannot have one without the other. It is this balance of rights and environmental needs – the interconnectedness of people to the natural world – that must be protected and enshrined in appropriate legal and policy instruments at national, international, and intertemporal scales because environmental rights are *emergent* and inherently future-facing. Modern environmental problems are rooted in human activity and will continue to emerge long into the future in often unexpected ways. Thus, we can consider our rights to a safe environment with clean water, air, and soil as emergent due to an interrelated "complex web of responsibility and liberty that includes our ancestors, as well as future persons, whose actions or welfare will be hugely affected by our decisions while we are alive" (Hiskes 2005, p. 1354). A final but crucial aspect of environmental rights is that it frames synergies over non-reciprocity between generational justice – for clean air, water, soil, and a stable environment now are in the interest of and benefit present and future generations (ibid.).

Thus, in complementary ways, sufficientarianism and environmental rights potentially resolve the issues of non-identity, non-existence, non-reciprocity, and discounting. They do this by both positing the collective social goods of environmental rights, specifically the soil, air, water, and ecosystems we rely on, and as we will see later, are fundamentally part of, while sufficientarian approaches stretch this into equally important and non-substitutable social good like peace and democracy. By positing transgenerational social goods, they sidestep the challenge that specific people do not yet exist in the future and the stickier arguments about who will exist depending on what we do. As Shue (2014) argues, we should assume that there will be future generations of some sort and we owe them consideration regardless of who they are. Implicitly, these combined approaches also resist discounting as maintaining certain ecosystem functions will likely require increasing investment and weight on the future, at least in the short term (Page 2007b). However, we also note that capability, sufficientarianism, and rights-based thinking still have airs of instrumentalism and anthropocentrism in the way they treat the environment (Winter 2022). Exploring the roots of *liberalism* that underpin these concepts will throw some light on how they might be adapted.

Ontological Challenges

The Cartesian dualism inherent to technocratic solutions to inter- and intra-generational justice is a problem. The ontology of division between subject and object has its roots in a dispute between Heraclitus and Parmenides over the nature of reality – the former saw the ostensible fixity of objects of perception as illusionary and instead posited continuous flux and change using the metaphor of a river – the latter argued change was the illusion and that objects must have some independent form that exists outside time (Graeber 2001). Following Parmenides, Plato (2000) in *The Republic* develops the idea of eternal forms that like the shadows from a fire on a cave wall are unclearly related to the objects but represent an ideal outside of time, as well as outlining his ideal class-based polity. Thus, Parmenides won the argument, and Western philosophy followed a track through which subject/object divisions abound, chief amongst these being society/nature (Hobbes 1914; Locke 2015) and mind/body (Descartes 2017). A problem with this ontological position as a foundation for justice is that it is fundamentally reductionist, defining materiality in terms of categories and boundaries, when our contemporary science and philosophy of the environment are much more focused on flows, relationality, complexity, wicked problems, and porous borders (Ginn et al. 2018; Haraway 2016; Quijano 2010). An irony with this is that without the bordering, measuring, and quantifying of the subject/object approach, we would likely have never realised it was not a good ontological model for the world (Graeber 2001). Thus, as Moore (2015; Fraser 2021) argues, we do not need to get rid of distinctions such as that between society/nature and mind/body, rather just recognise their dialectical nature and remain cognisant of their complex interactions.

A further but related issue is the individualism of Western philosophy, which can be seen as a logical assumption following from the subject–object binary. If objects are distinct and the mind can be distinguished from the body, then it is a small step to defining people as individuals over relational beings. Graeber and Wengrow (2021) describe the emergence of a curiously Western individual form of philosophical theorising that is unlike any other – a lone, white man sat thinking abstract thoughts – while in many other cultures this is a much more collective or at least social process – one that involves dialogue and shared understanding. Plato's work largely consisted of dialogues, and these collective and social forms of philosophy are found in many non-Western traditions (Kimmerer 2013; Sioui 1999; Watane 2022; Winter 2022). This way of doing philosophy as an individual clearly leads to positing the rest of human society as a series of individuals, most clearly and concisely captured in Descartes' *I think there I am.* This way of thinking has bled into and arguably dominates the most influential and consequential academic discipline, mainstream economics, and its object of analysis, the rational utility maximiser, *homo economicus*. This discipline, which has the ear of power, performs Bhaskar's (1997) epistemic fallacy – confusing how we know, measure, and demarcate things, with the question of whether those things actually exist beyond the measurer's mind. However, mainstream economics does this in much more applied and damaging ways than the philosophers who first conceived this epistemic fallacy. The fallacies of mainstream economics will be addressed in Chapter three.

A third problem, which again stems from the ontology of subject–object division, is a persistent anthropocentricism in moral philosophy. The separation of human subject from environmental object, with the former subordinate to the latter, has led us to believe that the totality of the environment can be imagined as a series of inputs and outputs – resources for extraction and sinks for pollution (Gelderloos 2022; Santos 2016; Merchant 1989). The consequences of this anthropocentric instrumental reason are unyielding environmental catastrophe. For instance, the crisis of biodiversity – what has been dubbed the sixth mass extinction (IPBES 2019) – is in turn significantly connected to agriculture and its use of land, and synthetic fertiliser and pesticides (Patel & Moore 2020). Animal farming uses vast amounts of land and food, which in turn requires land better used to feed humans, while also being a significant contributor to GHG emissions and a significant vector of infectious disease (Vettese & Pendergrass 2022). Most current biological science questions this society–nature binary. Moreover, current biological science characterises our very bodies are sites of multi-organism symbiosis, or we are holobionts, with any human a complex hive of bacteria, archaea, viruses, and eukaryotes, with increasing evidence that the gut microbiome regulates all kinds of human capacities (Gilbert 2020; Haraway 2016; Ogunrinola et al. 2020). Ecological science meanwhile shows that trees have complex, communicative interactions with and beyond their own species (Simard 2018), while the soil is teeming with an incredible diversity of organisms, layers, and micro-habitats, each playing a role in nutrient cycling and decomposition, and collectively maintaining the functioning of all terrestrial ecosystems.

Overall, there is a series of tragic ironies here. As Graeber (2001) argues, it is scientific reductionism and its limits that have shown that a narrow scientific, empirical ontology that separates and isolates objects is a poor model for complex, wicked problems, and more generally the Earth system. Similarly, it is through sciences that we know that the CEC is happening and, in many ways, how, with clear and significant uncertainties in this understanding. However, it is also technoscience and its capitalistic techno-fix, thinking and outputs like EVs, that drive the CEC by treating socio-political issues as engineering technical problems (Erickson 2016; Adams 2016). Simultaneously, technoscience promotes dubious and highly risky solutions such as geoengineering (Dryzek & Pickering 2019). The final twin ironies here are humans having the power to transform the world and imagining this world as inert and "without powers or transformations" (Graeber 2001, p. 51). This Promethean hubris leads some to favour transforming the Earth system, and still others to colonise space, rather than transform the social system. Thus, a more situated (Haraway 2016) post-normal science that accepts inherent uncertainties and systemic complexity (Funtowicz & Ravetz 1993) and further a more democratically mediated and responsive science and scientifically educated public are arguably needed (Erickson 2016).

Conclusion

Humanity faces a complex set of socio-ecological challenges. These challenges overlap, and their interactive nature makes them fundamentally *wicked* – difficult to define and difficult to reduce into manageable problems, soundbites, economic principles, or simple policy levers (Churchman 1967). There is a growing recognition amongst philosophers and social scientists that a reductionist, liberal, and Cartesian ontological model limits our capacity to think about interconnectedness between nature and culture, between human and more-than-human ecologies, and between current and future generations. Environmental philosophers expressed frustration at what seems like a pedantic insistence that future generations cannot be fully imagined and given an identity, and thus should be stripped of rights. If we accept the systemic complexity of environmental systems, the interaction between human needs and environmental quality, we must also extend this thinking of complexity and flow across time. Sufficientarian and capability thinking allows us to think in terms of thresholds, minimum standards, and human dignity. Combining these notions of minimum thresholds for human dignity with the environmental rights framework allows us to better imagine the lives of future people through the lens of current marginalised groups within society. We see how discrimination, institutionalised racism, socio-economic inequality, and unequal distribution of wealth and resources lead to a situation in which individuals lack the capability to meet their own needs. By understanding that capabilities, dignity, rights, and environmental resources are all connected, we can better understand how to imagine and integrate future generational interests as connected to our own. Thus, in the next chapter, we explore alternative philosophical traditions that emphasise connection, imagination, and practical action to resolve the challenges of intergenerational environmental justice.

2 Alternative Philosophical Traditions

This chapter traces an alternative philosophical tradition that subverts the Cartesian ontological divide of society/nature and mind/body and applies this thinking to intergenerational justice. Broadly situated within the discursive, epistemological, and ontological moves and perspectives of connection and dialectics, we endeavour to see the whole, open, and interacting nature of socio-environmental systems over time, while acknowledging that this can never be fully successful (Burns 2007; Graeber 2001). This endeavour involves being humble in accepting uncertainty. It also requires us to see flow and relationality within our world, and how these elements change over time as lens through which to understand intergenerational relations. This chapter will first visit a contemporary of Descartes, Spinoza, who was an early thinker that rejected Cartesian binaries. It will then briefly summarise Mauss' *The Gift* as a core text of intergenerational and intragenerational reciprocity; then it will consider insights of this work via Sahlins, Graeber, and Fritsch. Next, it will examine indigenous thinking around intra- and intergenerational justice. Finally, it will synthesise these strands of philosophy in a sketch of the human across deep time. The main argument being made in this chapter is that to achieve more substantive and inclusive forms of justice – now and in the future – requires new modes of thinking about the world and about ourselves. These modes of thinking can be and are drawn from the past, but they must be refashioned and reformulated for the future.

We can trace a contrary, minority philosophical tradition via a contemporary of Descartes, Baruch Spinoza. Spinoza took an overall relational position via a particular religious viewpoint – he advocated for a pantheistic conception of God, in which God was not a personal deity separate from the world but rather identical to the entirety of nature itself. Inherent to this philosophical tradition is the notion that God is immanent in all things – essentially forming the underlying unity of the universe itself. Spinoza (1982) argues against Cartesian distinctions between mind and body, people, and objects. He argued for a *monistic* view of reality in which there was only one substance in the universe, which he variously referred to as "God,"

DOI: 10.4324/9781003422686-4

or "Nature." This substance is infinite, eternal, and self-caused; to Spinoza, everything that exists is an expression or mode of this single substance; thus, interconnections between nature and culture are inherent. Hickel (2020) traces a line via Spinoza through phenomenology that shows all we know/experience is through interaction and interrelation. Thus, these ideas can be related to the Greek dispute between Heraclitus and Parmenides, as Graeber (2001) points to, but also Fritsch's (2018) argument that people from whatever generations are co-constituted by both past generations and their ecosystems and the Earth. This position of seeing relationality over separation or binaries emerges in the progressive anthropological literature considering the role of reciprocity in social relations and is discussed in further detail in this chapter.

Of note in our analysis is the work of Gilles Deleuze and Felix Guattari. Deleuze (1990), clearly influenced by Spinoza, developed complex and innovative metaphysics, ontology, and epistemology. Deleuze articulates the concept of emanation (as opposed to creation) – that there is a slow manifestation of lower forms of existence from higher or more unified levels of reality. The metaphysics of emanation implies a hierarchical structure, for which the ultimate source or principle of existence gives rise to successive levels of being. In essence, God did not create people as distinct from itself, but rather tore part of its body – thus, humanity and the world emanate relationally from a source. This metaphysical tradition challenges traditional notions of hierarchy and transcendent origins. To Deleuze and Guattari (1987), the concept of rhizomatic thought is essential – it is a metaphorical framework for understanding the structure of knowledge, culture, and social organisation as being interconnected, decentralised, and nonlinear. Rhizomatic (a metaphor taken from certain root structures such as those found in bamboo, e.g.) links to the idea of spreading horizontally underground, creating networks of interconnected nodes, without a central trunk or hierarchical system. In the context of philosophy, rhizomatic thinking emphasises networks of interconnected elements without a fixed centre or higher operating principle. The metaphorical framework of rhizomatic thinking stresses how a multiplicity of connections and relationships between different elements provides a dynamic network of significance – one that is decentralised and conceptually "nomadic" – it allows for fluid movement, adaptation, and change across different contexts and territories. A rhizomatic thinker traverses different landscapes of knowledge and experience, without being confined to a single ontological position or philosophical perspective. In contrast to the positions emphasised in the previous chapter, rhizomatic thought invites us to challenge traditional hierarchies, binaries, and linearity, and encourages us to express creativity in our thinking; thus, we embrace dynamic networks of ideas, remain open to multiple interpretations, and are inclusive of different voices (ibid.).

Social Relations of the Gift

Marcel Mauss was a French sociologist and an anthropologist known for the influential work *The Gift: the Forms and Functions of Exchange in Archaic Societies*, published in 1925. *The Gift* (2002) asks two questions:

> *What role of legality and self-interest, in societies of a backward or archaic type, compels the gift that has been received to be obligatorily reciprocated? What power resides in the object given that causes its recipient to pay it back?*

Mauss explores the phenomenon of gift giving in traditional societies, particularly focusing on those within Polynesian cultures, but drawing examples from across the world. He argues that gift exchange is not simply a matter of economic transaction. Rather, it carries much deeper, social, cultural, and symbolic meanings for both the givers and the receivers of gifts. Mauss surveyed anthropological work on gift exchange with an eye to crises in his era's legal and economic organisation, thus exploring "questions that are old but ever-new" (Mauss 2002, pp. 4–5). Mauss explores how normative social relations emerge, suggesting these first appear as gifts and sacrifices to the gods, who as the lords of air, sea, and earth gave so much and asked for little, but if refused could wreak havoc. This unequal, symbolic, and shifting reciprocity becomes for Mauss the basis of sociability, as "everything passes to and fro" in order to balance accounts, and ultimately keep the peace (Mauss 2002, p. 17). The ideas that Mauss explores in this work concerning the role of gifts as a form of symbolism and prestige – how gifts are laden with symbolic meanings that are established within certain social hierarchies within communities. A gift is determined and defined not only by its material value but also by the social meanings and obligations that are attached to it by the culture at large. Generosity is thus a highly valued trait in many societies: for example, Mauss discusses the Potlatch ceremony amongst Indigenous peoples of the Pacific Northwest coast. In the ceremony, individuals engage in highly competitive gift giving and lavish displays of wealth. The ceremony is an expression of a complex social and ritual event which serves to reinforce hierarchies of social status, but also redistributes wealth and affirms communal bonds between different groups with varying social status amongst them. Gift giving is thus the means to foster social cohesion and cultural solidarity within society, and as discussed in the previous chapter, this principle of reciprocity reinforces social relationships, alliances, and networks of mutual support. Gifts are a means of building and maintaining social position amongst a group, so the symbolic value is deeply significant. Mauss discussed how gift exchange can be construed as a social practice that challenges conventional classical economic models that define the value of something solely through rational self-interest and market exchange. The logic of the gift operates on a completely different set of principles, emphasising the role of social obligation and reciprocity as overriding rational self-interest as defined through the minimisation of losses and the maximisation of gains to the individual (all from Bird-David & Darr 2009; Frank 2016; Pyyhtien 2016).

The conclusions Mauss (2002) makes are threefold. Firstly, he posits that the objects and things given embody a part of the spirit of the giver, thus leaving both a duty of return and an almost fetishised will in the object to return it to the giver. Secondly, he posits that reciprocity-based social relations embodied in the gift – broadly as the Māori proverb goes, to not take more than you give – were the prior stage to state-based and market-based societies, as can be seen in many surviving social customs. This produces an alienating tension as the reciprocity-based systems are overwritten by the "abstraction and inhumanity" modern law and economics. Thirdly, he argues that for the present society to move forward, it must go back to "archaic society and elements of it," namely "solicitude arising from reciprocity and cooperation" and people with an eye to both their own and society's interest (Mauss 2002, pp. 86–89). This essay has been the subject of much debate since and has relevance to not only inter- and intragenerational justice but also science fiction speculation. Kim Stanley Robinson in his *Mars Trilogy* (1992, 1993, 1996) has his Mars colonists, who rebel against the Earth's capitalism and colonialism, explore new forms of democracy and economics. Part of this new economics is the reintroduction of the social relations of gift exchange. This can be seen as both an answer to Mauss' call and a challenge to the fading vision of capitalist notions of progress and linearity.

Placing *The Gift* firmly and fairly in the tradition of Hobbes and Rousseau, Sahlins (2017; Graeber 2001) argues that Mauss was trying to explain how order emerges from disorder. However, while asking a similar question to Hobbes, the answers given could not be more different – Hobbes answers with the vertical state, monarch, or other such overlord that takes on authority and the monopoly of violence to ensure peace. In contrast, Mauss offers something horizontal, a between relation that suspends force by will, custom, and ritual via the reciprocity of the gift. Sahlins (2017) expands this comparison between Hobbes and Mauss but concludes, as Mauss concludes, that the fine line between war/conflict and trade/festivity is fragile and that we can never really, and should not try, to separate the material and economic from the social.

Graeber (2001, p. 154) agrees with Sahlins that the "ghost of Hobbes lingers" over this account. He adds several interesting observations. Firstly, contrary to modern neoliberal thinking that would categorically distinguish relations of violence and economic self-interest, Graeber argues these are variations of a theme – not caring about other people and their fate. Secondly, he puts Mauss in a longer context than Sahlins, placing this in a 2,000-year-old argument about how any gift giving had self-interest attached, positing a specious distinction between self-interest and altruism that Mauss' essay so skilfully blurs. He adds that despite its limitations from a Marxian angle, the essay does add to the theory of alienation, showing that like socialisation not ending on leaving school, nor does the creation of objects end on the factory floor – "things are continually being maintained, altered, and above all, vested in new meanings, even as they are often repeatedly detached and alienated again" (Graeber 2001, p. 163). Finally, Graeber (2001) laments the underlying assumption that it is peace and order that we must explain/question, not the potential for violence and conflict. These assumptions not only overdetermine human nature as essentially violent and prone to competition and conflict, as Hobbes does, but also ultimately help justify the present's police and prison industrial complex (Vergès 2019, 2020; Butler 2020). Intergenerational justice concerns the human and the more-than-human, but within the former there is a conflict that spans deep time – what is the human and what is its nature, and is this one of being or becoming? These questions will be returned to in the final part of this chapter.

Indigenous Perspectives on Justice and Time

In many ways, the turn to indigenous knowledge by interests within the UN (Watane 2022) or the IPCC (2022) is a testament to the durability of these ideas in the face of terrible oppression. It also speaks of the strength of circular notions of time (Estes 2019; Sioui 1999), which it is argued here offer more hope of transgenerational solidarity and responsibility. This is not in the crude sense that denies the linearity of time, but in the sense of natural cycles of season, day and night, birth then youth then eventually death, and then it starts again. This can be an antidote to the Western myth of human and social progress that is tied to a linear notion of time but loads this with socio-cultural baggage. This baggage presents the spoken and unspoken – modernity, democracy, technology, and social progress, darkly mirrored by oppression, racial discrimination, and concentration of power by an ethnoclass elite (Mignolo 2007; Wynter 2003). This section will explore how indigenous societies frame justice in both transgenerational and more-than-human ways, thus offering practical, pragmatic, and normative solutions.

Indigenous knowledge and indigenous philosophy refer to the complex body of practices, skills, beliefs, and experiences that are developed and transmitted over generations within specific cultural groups or communities. The scope of indigenous knowledge is extremely broad: covering the observations and understanding of natural resources, environmental health, ritual, spirituality, and social organisation, each unique to the culture from which it is derived. Indigenous knowledge is inherently *local*, in the sense that it is tied to understanding and ecological relationships connected to places but is nevertheless sustained over very long periods of time. Given the complexity and scope of indigenous knowledge, broad generalisations are difficult and risk creating new epistemic injustices by taking such knowledge out of context and applying it without consent. Nevertheless, many indigenous thinkers have presented a pattern of epistemologies that frame humans as fundamentally relational in space, time, and ecology (Winter 2022; Watane 2022; Estes 2019; Kimmerer 2013), giving rise to growing recognition of the value of indigenous knowledge systems, not only for the communities that they have served for centuries but also from the wisdom that such approaches bring to communities outside of that culture. Indigenous knowledge is vital both for preserving cultural heritage and in addressing contemporary global environmental catastrophe (Santos 2016; Green and Raygorodetsky 2010; Petzold et al. 2020). The following section will try to tentatively summarise some common features of indigenous philosophy that shed light on new ways of thinking about our world and ourselves that offer more substantive and inclusive forms of justice.

Kimmerer (2013, p. 78) details how she works on her garden pond, and it works on her as a eutrophic woman who tends a "circle of care . . . [T]here are grandchildren to nurture, and frog children, nestlings, goslings, seedlings, and spores." This is framed as a democracy of species embedded in and relying on a water net connecting us all with a cyclical temporality of life leading to death leading to life (Kimmerer 2013). More comprehensively, Engle et al. (2022; Winter 2022) propose accountability to all peoples, future generations, and the Earth. Firstly, they (Engle et al. 2022, p. 8; Winter 2022) argue that current liberal democracies are accountable to a small subset of people and special interests, while others like "children, migrant workers, and people in prison have little to no say," and that "obscene inequalities within and between cities and countries" must be "acknowledged and acted on" if we are to "heal and move forward as a species." Secondly, they claim that the responsibility to posterity that was enshrined in the origins of democracy has been replaced by a "hegemony of short-termism . . . political election cycles and quarterly corporate logics." To be accountable to future generations, we need "long termism in policy, funding, and investment decisions," alongside new forms of "engagement and proxies for accountability" that can better represent posterity. Finally, while many constitutions and legal systems acknowledge rights of people and corporations, indigenous governments and philosophy promote the "inherent rights of nature." Such a perspective demands living in a "respectful and reciprocal relationship with the Earth and having accountability for healthy ecosystems," which will be essential in dealing with the CEC, biodiversity loss, and remaining within the Earth planetary boundaries (Engle et al. 2022; Winter 2022).

Sioui (1999), writing a history of his Wendat people, reveals a more democratic, nuanced, and complex society, questioning ideas of primitive, pre-contact societies, but also more broadly the idea of social evolution. Describing the Wendats as a Circle Society, he affirms that this is in the sense of temporal cycles and interconnections with the animate and non-animate more-than-human environment we are entangled in. However, this Circle Society also refers to a complex social system including communally held land; leadership positions demanding much and giving little yet still controlled and balanced in multiple ways; a division of labour between genders that allowed for women's independence and power, with women overseeing internal community issues and men overseeing intra-community issues, but this divide complicated in multiple ways; and a complex system of clans that mixed family groups and adoptees from beyond, building social solidarity and responsibility beyond blood ties (Sioui 1999). Overall, this history and argument present significant challenges to social sciences. Social sciences still present a linear social history from simple to complex, from hunter gatherer, via agriculture to industrialisation and urbanisation, with the benefits of modernity and complexity a package deal along with the costs (Graeber & Wengrow 2021; Mignolo 2007). However, the summary above suggests pre-contact societies were not so primitive and display complex forms of ecological and social understanding. This is socioecological knowledge that we, from the imperial core, are having to relearn and struggle for. This is a contradiction that makes the notion of Western progress difficult to maintain, especially considering the many wicked problems this *progress* has produced.

Finally, on tying intragenerational and intergenerational justice together, indigenous theory and philosophy are explicit. As stated earlier, Kimmerer (2013) sees and argues for a democracy of species co-existing across time and space. Watane (2022) details how this works in practice from a Māori perspective. This involves "face-to-face" encounters with ancestors, leading to facing each other in the present, then this encounter allowing the pushing of the moral imagination into the future. Watane (2022) describes how facing ancestors is both linguistically and narratively evident in indigenous societies. Linguistically, Táíwò (2022) points out how the Yoruba pronoun for older people refers to a group – suggesting this includes past people and historical accumulations. Similarly, but more expansive, thc Maori inquiry into somebody's identity translates as "who are your waters?," placing persons in complex webs of socioecological and temporal relations. Narratively, Polynesian stories of sea faring and tales of long distances travelled over land "weave connections of generations and land and waterways" that enable and require the recognition and moral imagining of an intergenerational community "far beyond our own temporally and spatially-bound lives" (Watane 2022, p. 6). Facing the past in this way, we see how ancestors collectively shaped the present and how we do the same now, but in Marx's (1852) sense of not how we would choose to. By facing each other in the present, we see not only how structural injustice in the past and present shape harmful interactions, but also how cooperation and caring offer the potential for new beginnings and healing. Finally, this leads to facing the future, acting like in, Táíwò's words (2022; Krishnan 2022), ancestors, in acknowledgement of "the way we each embody collective memories, hopes, and aspirations for the future. Just as our ancestors did, we too hope for a world that we know we will never live to see ourselves" (Watane 2022, p. 7). Table 2.1 summarises these contrasts between liberal and indigenous philosophies.

Thus, indigenous knowledge offers much but also asks our world much. It sees beyond the Cartesian binaries that make human society exceptional to the world with all the problems this fallacy creates. In accepting of our place in the land, perhaps we can move beyond the instrumental treatment of our ecosystems as a series of inputs and outputs. This is important as there is a danger in the mitigation of the climate crisis to treat this with an almost fetishised focus on CO_2 alone, ignoring or downplaying the multifaceted but connected problems (Vettese & Pendergrass 2022; Hickel 2020). Strong sustainability or the sufficientarian position would argue that certain things like biodiversity cannot be replaced, and these socio-environmental goods are in many cases essential aspects of ecosystem functioning and thus people's lives. Indigenous voices such as Kimmerer (2013) would add this is simply the right and ethical thing to do in a democracy of species. Indigenous knowledge offers a renewal and revitalisation of our sclerotic and attenuating democracies, showing these as social institutions that can and must adapt and whose role is to ensure justice and well-being, not corporate profits, security, and economic growth (Gilmore 2023; Engle et al. 2022). A major aspect of this democratisation is the tying of intra- to intergenerational justice, which is implicit in the respect for, and cognisance of, humanity's place entangled in the land. This land is the source of all human livelihoods ultimately and, if sustained, will continue to be so.

However, the costs to a neoliberal political and economic system are heavy and potentially fatal. It requires remembering and an accounting with the past that current presenteeism and the institutional and discursive powers of strategic ignorance and forgetting resist (Táíwò 2022; McGoey 2019; Patel & Moore 2020). The capitalist system creates commodities quite clearly via factories, for example, but less obviously it also creates people via systems of care, schooling, and socialisation; however, the market that connects people and products also acts as a vast occlusion mechanism rendering the production of people and things invisible (Graeber 2001). Thus, part of overcoming these forces of occlusion involves looking into the past (Watane 2022, p. 10):

> [T]o find value in understanding how opportunities and challenges were navigated previously, to reframe and grow existing solutions in different ways, and to retrace our steps in search of opportunities for change.

This involves recognition of how much of today's wealth and injustice can be traced from current exploitation to a longer arch of neocolonial extraction and ultimately to earlier periods of colonialism (Gilmore 2023; Táíwò 2022; Hickel 2020; Patel & Moore 2020; Shue 2014). Today most of the people

Table 2.1 Cartesian and Rationalist Worldviews Versus Indigenous and Relational Worldviews

Cartesian/Rationalist Worldview	*Indigenous and Relational Worldview*
Material Property-based • Property ownership is a fundamental right • Anything not human can be property • The right to extract value from property • Society and politics aimed at growth and excess production • Land, water, and sea are measurable/quantifiable property • Property has a monetary value and may be transferred and speculated upon	***Relational Place-based*** • Relationship to land, water, flora, and fauna core to identity • "Ownership" of land, water, flora, and fauna, not individual right or social concept • Territory is marked and described but in collective sense not individual • Social reproduction focuses on sufficiency, not excess • Land, water, and sea are culturally embedded in reciprocal relations to which philosophy and law emerge • Culture and place are entwined, and so if place is degraded or appropriated, the culture is also damaged
Individualistic • The individual and their autonomy/freedom is the ultimate value • In communal conflicts with individuals, often the individual is privileged • Society's role is to not overly constrain the individual's freedom	***Collectivist*** • The autonomy of the individual comes from the relative strength of the community • Solutions to tensions in community found through the community and consensus • The collective has responsibility to support the individual
Anthropocentric • Non-human nature is treated instrumentally – i.e. what can it do for humanity • Humans privileged as dominant being, with justice and dignity concerning only humans • A strong separation between the cultural and the environmental with the former seen as paramount	***Cosmological*** • Humans not privileged but have specific duties as part of a multi-being community • Humans part of a wider whole, and justice and dignity for humanity demand justice for the wider whole • Culture and environment not distinguished as culture comes from the encounter with land and the beings in it – nature is everywhere in everything
Discontinuous Temporality • Present time is seen as most important time • The past exists but is not as important as the present • The future is considered, if discounted, but distant future is almost completely ignored in decision-making • Time is seen as a linear line	***Continuous Temporality*** • The present is not privileged • Past and future seen as integral to the present • Time seen as circular or a spiral like the seasons, allowing for the past and future to "speak" to the present

Source: Adapted from Winter (2022)

facing the worst of the CEC, both within and between countries, are descendants of the ethnoclass that was colonised (Paradies 2020; Quijano 2010). Therefore, this ethically requires some form of redress, or reparations as part of a just transition (Táíwò 2022). Using indigenous knowledge without this redress and justice is a form of and continuation of epistemic violence (Winter 2022). A further cost to this neoliberal system would be a renewed and transformed notion of care. Currently, care is anthropocentric, highly financialised, extractive, highly racialised, and gendered, with a shadow system of care provided disproportionately by women for free. However, if this care was removed or withheld, the world economy would crumble (Fraser 2021; Vergès 2019). It is unlikely a neoliberalism that was not anthropocentric, remembered and learned from history in a progressive and justice-centric way, and had an expanded and justice-centred notion of care would be neoliberal or even capitalist. However, these appear to be the conditions of indigenous philosophy and perhaps the hopes for a more justice-focused present and future.

Defining the Human Across Deep Time

We can perhaps begin with the way literary theory began to question the power of the author over textual interpretation. Barthes (1977) claims that although the author remains powerful, many had shown this power to be only to limit meaning or to impose a narrow meaning that language itself sought to escape. He goes on to claim that what we "know now" is that rather than a unified, theological meaning of the Author-God, the text is a "multi-dimensional space in which a variety of writings, none of them original, blend and clash" (Barthes 1977, p. 146). Elaborating on these ideas, Foucault (1994, p. 215) talks of the *author function* that, in some texts and some discourses, gives authority to an author to define under a "principle of a certain unity" the text. This unity instead of producing a wealth of meaning, similar to capitalism, creates a false scarcity and limitation of signification. Crucially, this unity of authorial intent and consistency is based upon a type of Christian exegesis, which in effect judges this authority/authorship based on internal coherence (Foucault 1994, p. 214). As Foucault's thought moved from text to the subject, this idea of the author function begins to be applied to human subjectivity. Historically, it was the Enlightenment, Foucault (1996, pp. 312–313) contends, that "simultaneously problematises man's [*sic*] relation to the present, man's historical mode of being, and the constitution of the self as an autonomous subject." As such, we must "proceed with the analysis of ourselves as beings who are historically determined, to a certain extent, by the Enlightenment." He goes on to say that critique should no longer be searching for human universal values but rather a historicised investigation into "the events that have led us to constitute ourselves and to recognise ourselves as subjects of what we are doing, thinking, saying" (Foucault 1996, p. 315). Thus, both Foucault and Barthes, in undermining this author function and offering an "inherently pluralist" and differentiated text, were pointing towards the human subject and what it would take for this figure to understand that it writes itself (Dean & Zamora 2021).

The Overdetermination of Man

This is where we may depart from Foucault, as he favoured an individualised form of experimentation of the self and saw something liberatory within an abstract form of neoliberalism that has proven largely chimeric (Dean & Zamora 2021). More interesting and aligned with this book's arguments are decolonial thinkers such as Sylvia Wynter (2003) and Walter Mignolo (2007), who provide the theoretical support; indigenous writers such as Sioui (1999), Estes (2019), Whyte (2017), Kimmerer (2013), and Mann (2001), who provide a praxis of resistance to coloniality/capitalism and historic examples of alternatives; and finally Graeber & Wengrow (2021), who collate evidence from these indigenous scholars and the deeper archaeological human record. Collating these ideas, evidence and history provide a new/old human subjectivity that is both freer than today's and more associated with the deep historical record. It is to Graeber & Wengrow's (2021) arguments we now turn.

Of all the determinisms or essentialisations we are offered by capitalism, that of human *nature* is arguably the most damaging and limiting. Graeber and Wengrow (2021), in concert with Foucault, locate a source of this determinism in the Enlightenment. They trace two dominant stories that frame this nature in quite different ways but that have similar damaging and limiting effects upon the present. The first story is a rehashing of the biblical story – we were born in a state of grace, we messed up, and now we have great cultural achievements, mirrored by the horrors of the colony, prison, police, and war. The more secular version comes from Rousseau, and instead of original sin, we were free and equal; then institute agriculture and private property and it all goes downhill from there. The other story is more familiar and comes largely from Hobbes' *Leviathan* and shortcuts the first story by saying primitive life was brutish, short, and mean, as humans are inherently violent and selfish. This changed with the emergence of authority in the state/sovereign that imposed limits and rules, in turn allowing the fruits of civilisation to flourish. While saying quite different things, both stories create problems today that this book grapples with. Firstly, both imply evolution from simple to complex, societies of small numbers evolving to large, and as we move up this ladder, we get both the cultural fruits and the oppressive cages. The fall from grace version is a problem as it points to some distant time where things were *better* and then assumes that all is lost as private property and capitalism are here to stay. This leaves us to focus on working at smoothing the edges of inequality while leaving the structural conditions of injustice inherent to the colony and capitalism intact and untroubled. The other story is more damaging as at its core is the figure of *Homo economicus*, the rational and selfish utility maximiser, thrust back in time (Sahlins 2017), who requires the state and its oppressive apparatus to control the game of accumulation that would descend into violence otherwise (Graeber & Wengrow 2021). Thus, both stories limit human potential and leave unquestioned the structural factors that frame and reproduce such limited subjectivities and injustice.

Another (Graeber & Wengrow 2021), and in many ways more important, problem is that both these stories have little basis in anthropological or archaeological records. This record is instead much more interesting, kaleidoscopic, and exciting, with scarcity largely part of the modern neoliberal/capitalist condition being projected backwards rather than a general Neolithic condition (Sahlins 2017; Sioui 1999); highly hierarchical societies bordering highly egalitarian societies; large and diffuse cultural areas where an individual could travel many hundreds of miles and still find shelter and welcome (Sioui 1999); city states and more mobile societies working and living in none hierarchical modes, or with matriarchies and/or matrilinear organisation (Saini 2023; Powell 2018; Voyles 2015); societies using flood retreat and related techniques of *play*farming, meaning much of the work is diminished and showing cognisance of these dynamics, or societies trying farming, then returning to hunter-gathering and so refuting the idea of linear evolution; and flexible polities involving more or less authority and hierarchies depending on environmental and seasonal conditions. Overall, this empirically detailed book deconstructs these two *just so* stories about our human *nature*. Instead, it offers a much richer and open visa of human possibility in which humans from village and city to wider geographical areas, consciously and experimentally adapted and changed their socio-political organisations (Graeber & Wengrow 2021).

In support of this radical revision of human history, Whyte (2017) describes his own Potawatomi people and how they lived before the colonial encounter. This included living and roaming over 30 million acres, seeing themselves as part of a multispecies ecosystem and steward to animals and plants within this area, and having a "long tradition" of a seasonal round in which the "purpose, organisation and size of cultural and political institutions (from ceremonies to villages to bands) changed throughout the year depending on what plants and animals needed to be harvested, monitored, stored or honoured" (Whyte 2017, p. 89). Regarding agriculture, Sioui (1999) claims that Wendat societies that farmed did not consider themselves as more advanced than their hunter-gatherer neighbours. Thus, from pre-colonial encounter societies in the Americas and throughout the archaeological record, we see that the human *nature* presented to us by both left and right traditions is merely a narrow slice of human potential. Lastly, this section will consider what decolonial discourse has to say upon the human subject for it bears many parallels with the above arguments.

A general strand of decolonial thinking seeks to decolonise *knowledge*, while a more specific strand seeks to deconstruct the overdetermined capitalist subjectivity of *man.* Mignolo (2007) labels the first strand *epistemic unlinking*, which entails a rejection of the *package* of democracy, modernity, and modernisation, as these are constituted by the occluded aspects of the colony/coloniality. However, in this rejection, one becomes subject "to all kinds of direct and indirect violence" (Mignolo 2007, p. 450). An example of this violence is provided by Mann (2017) when she reprints the FBI's surveillance documents of Indigenous activist/academic Ward Churchill. A more general aspect of this epistemic violence is the way the scientific research and specifically anthropology have subjected Indigenous people to some of the worst aspects of colonialism (Smith 1999), while at the same time relegating the colonial other to the *wrong* side of the Cartesian binaries of society/nature, civilised/savage, and rational/irrational (Santos 2016; Wynter 2003). Another aim here would be confirming the colonial other as savage and the aggressor as many cultural products did and do. Thus, decoloniality partly refers to freeing our minds, knowledge, and imaginations/imaginaries from the Eurocentric Totality. This totality "negates, exclude, occlude the difference and the possibilities of other totalities," so a key aspect of epistemic unlinking is the analytic uncovering of "silenced histories, repressed subjectivities, subalternised knowledges and language" all occluded under the Totality of Eurocentric Modernity/Rationality (Mignolo 2007, p. 451).

Focusing on decolonial subjectivities, Wynter (2003) traces how the *descriptive statement* of the human develops from pre-modern Christianity to modernity/coloniality. Moving from Christianity to the Enlightenment, the *descriptive statement* of the human mutates from a theocentric cosmology in which humans, sinful by nature, gain redemption through the church, to a humanist, secular version where the human is a rational political subject of the state and gains/proves his *reason/rationality* by adherence to state law. However, this is not a clean break from one description to another, for the Christian Others or Muslim or Jew are replaced by the colonial others of the world, signified by the colour line. Wynter (2003) argues that this colour line and race, while having no basis in biology, become the central organising principle of domination for the last 500 years. Across these years, the colonised move through distinct stages of spiritual, political, to contemporary economic damnation. *Man* is "now defined as a jobholding Breadwinner, and even more optimally, as a successful 'masterer of Natural Scarcity,'" while "Human Otherness [is] to be peopled by a new category, one now comprised of the jobless, the homeless, the Poor, the systemically made jobless and criminalised – of the 'underdeveloped' – all as the category of the economically" damned (Wynter 2003, p. 321), or subjects of Gilmore's (2023) organised abandonment. All the while, the knowledge that we humans are both the origin and definers of ourselves is repressed (Godelier 1999). This repression takes, firstly, a supranatural form with God as the creator/writer of what is human, only for a secular scientific, evolutionary *descriptive statement* to replace this. This replacement supports Wynter's (2003) general point that in the *invention of man*, there are, contrary to Foucault, many continuities and discontinuities. It also supports Feyerabend's (1977) argument that science has taken on many aspects of religion, not least in trying to overdetermine and limit human subjectivity. Thus, Wynter (2003, p. 268) argues that "one cannot 'unsettle' the 'coloniality of power' without a redescription of the human outside the terms of our present descriptive statement of the human, *Man*, and its over representation."

Conclusions: A New/Old Subjectivity for Intergenerational Justice

Overall, this poses the question of what this decolonial thinking, empirical deep history, and indigenous historiography offer in this book. From their survey of pre-colonial and deep past, Graeber and Wengrow (2021) formulate three general forms of freedom that could summarise what some societies of the past enjoyed, and that would make social democrats of today blush. Therefore, in the spirit of rejecting the myth of progress, this book advocates these three basic freedoms enjoyed in the past as something to aspire to in the future. These freedoms are (1) the freedom to move/relocate from one's locale; (2) the freedom to ignore or disobey commands of others; and (3) the freedom to shape different social realities, and to build and deconstruct social formations (Graeber & Wengrow 2021, p. 503). The irony here is that these freedoms are enjoyed by a small percentage of the world's population at the expense of the rest of us, which brings in a fundamental caveat. This is that these freedoms become meaningless if they harm others as this just descends into domination. It is no coincidence that societies of the past appeared to understand this and *actively* designed social institutions to counter the creep of authoritarian power (Boehm 1999). For instance, many societies required the chief to share all their wealth/resources, while the idea of leadership itself was alien to that today, in that it was a role of *service* with a function to *persuade* rather than *dictate* (Graeber & Wengrow 2021; Sioui 1999). Thus, the question becomes how we escape the modern pseudo democratic malaise. A malaise which seems to be getting worse with creeping authoritarianism to add to its performative aspects.

This comes to the old/new subjectivity that this book proposes, but which presents contradictions and dilemmas. It is contradictory in that how do we become relational, autonomous beings that understand our place within an ecosystem we depend upon? Does a human need to be autonomous before they realise or write/define themselves? The dilemmas relate to the frustration about the general lack of real progress in mitigating the CEC. Malm (2021, p. 8), detailing the fact that we are not progressing but regressing in CEC terms, asks "At what point do we escalate?" Sovacool and Dunlap (2022), similarly noting the failure of the global system to deal with the CEC, review forms of resistance from the passive to the actively violent. These thinkers do not advocate personal violence, but they do show that violence is a highly fluid term that is welded by the powerful against the weak (see also Butler 2020). However, it appears that breaking things, blocking, and generally doing things that will be *the investment risk* look increasingly essential to prevent climate breakdown (Mueller 2021; Malm 2021; Sovacool & Dunlap 2022). Thus, now as in the past, liberty and respect must be not only argued for and debated but also materially and bodily struggled for (Estes 2019). Therefore, this new/old subjectivity in being relationally autonomous must stand up and resist the injustices of our global order in whatever way they feel they can.

Finally, in the spirit of not making the mistake of overdetermining this subjectivity, we might conclude with a basic sketch: this new/old subjectivity will be relationally autonomous, which includes the cognisance of more-than-human relations; this figure will understand that it writes/defines itself within the bounds of the above relationality; this figure will understand that material artefacts such as technologies, money/debt, or processes such as using energy involve social relations that must explored/decolonised; finally, this figure will understand that language and knowledge themselves are forms of power and that these too must be carefully explored/decolonised. Again, this is nothing new as Graeber and Wengrow (2021), Boehm (1999), Malm (2016a), or Mueller (2021) argue and Sioui (1999), Estes (2019), and Whyte (2017) show. Fundamental to our contemporary world, the past will inform our future for good or ill.

3 Mainstream Economics and Scarce Justice

This chapter looks at what mainstream economics has to say about intergenerational justice and the CEC. As Fritsch (2018) points out, many economists have used Mauss' work on reciprocity to consider intergenerational justice with various results, but generally missing the point Mauss was making that market-based relations entail inhumane trade-offs in the justice sense. These economistic accounts of intergenerational justice will be considered first. Next, we will consider what we call the economics of the Anthropocene. This is a form of knowledge that has grown and colonised the world of knowledge parallel to how the capitalist economy has grown and colonised the world. However, both the economic knowledge and the economy it ostensibly describes are unsuitable for the world for scientific and social reasons. Thus, in keeping with the previous chapter, we require new forms of knowledge and thinking. This follows the simple principle that the ways of thinking, acting, and being that get you into a crisis are not likely to get you out of that crisis.

Generational Wealth Transfers

One strand of economic literature explores reciprocal relations over generations, focusing solely on these relations within the family. For example, Becker (1988) considers family finances, the motivations behind transfers within these families, and the consequences of this on macroeconomics and economic growth. The concern in this chapter is on "good" growth – in which the economy grows perpetually with declining fertility and increased investment in human capital and knowledge, against "bad" growth – which involves Malthusian growth in population and a declining share of scarce resources. Becker's (1988, p. 4) answer is a strange version of altruism (not the only motivations he admits) that parents bequest financially to children as units and which would be "negatively related" to the number of children, thus limiting population and investing better in human capital and knowledge. Arrondel and Masson (2006) examine the patterns and determinants of financial transfers between generations (i.e. bequests and inheritances) and inter vivos transfers occurring during the donor's lifetime (investment in tuition, family loans,

DOI: 10.4324/9781003422686-5

etc.), in shaping intergenerational wealth dynamics. Their analysis explores the patterns of cultural norms, inheritance laws, and parental preferences that influence the timing and size of such transfers. They refer directly to Mauss's work (2002) and his point that *the gift* is always ambivalent with the positive side of reciprocity in gift giving countered by the negative side of the debt incurred and the power this allows (see also Graeber 2011). They go on to suggest that while the primary role of parent-to-child gifts should be wealth transfer across generation, this is complicated by familial conflict with domineering and controlling parents and ungrateful children (Arrondel & Masson 2006). However, what neither of these arguments consider is that intrafamily wealth transfer is one of the many problems with neoliberal economics. Also, the tensions within the family model perhaps point to where oppression and domination emerge in deep history.

Familial dynasties, nepotism, and power relationships perpetuate wealth and human capital inequality across generations, and these dynamics have particularly affected the modern era. Piketty (2013, 2020) argues and shows that wealth and its intrafamily transfer are a fundamental factor in the vertiginous rise in inequality/injustice during the 20th and early 21st centuries. Thus, Becker's concept that there should be more investment in human capital to ensure future cycles of economic growth seems elusive. We live instead in a world of rentier empires, land, natural resources, and virtual real estate, holding monopoly and monopsony power that both drive injustice and inequality and stifle economic growth (Greene 2022; Christophers 2020). Fast-forwarding to speculative science fiction, it is not hard to extrapolate this into intergalactic dynasties where long-lived patriarchs populate and control whole planets and systems (Hamilton 2004, 2005).

While the focus on the family may miss the point of Mauss' work discussed in the previous chapter, which looked beyond kin relations, it may touch on how and where human domineering relations emerge. In his early work, Graeber (2001) ruminates on where unjust relations emerge and suggests that these may start in the home. In surveying the archaeological and anthropological evidence of deep human history, Graeber and Wengrow (2021) look at the evidence. They find a kaleidoscopic variety of social relations and conscious manipulation of these relations and systems, a variety and fluidity of social organisation that we seem to have forgotten in liberal democracies (Graeber & Wengrow 2021).

If we have choices and we can have freer societies characterised by solidarity and cooperation as much as conflict and oppression, it seems, contrary to Hobbes and Mauss, we must question where and how oppressive relations emerge. Drawing on the tragic life of Franz Steiner, Graeber and Wengrow (2021) speculate that oppression emerges from the breakdown of the first form of freedom discussed at the end of the last chapter, the freedom to move, and from forms of charity. The question of how stateless egalitarian bands become top-down chiefdoms and states is perhaps a process whereby a nominal leader accepts into their household widows, orphans, and all sorts of misfits and war captives. This growing household becomes a locus of power within and without, with a patriarchal authority imposing its will – an aspect of the negative side of Mauss' gift of charity – on those within the household, and a host of henchmen and administrators to do the oppression without the household (Graeber & Wengrow 2021). Whether this is a source or the source of patriarchal relations is perhaps difficult to answer. However, speculating and collating vast amounts of evidence, as Graeber and Wengrow (2021) do, is better than essentialising a fixed human nature without recourse to evidence as mainstream economics does. Again, perhaps looking here at speculative science fiction is helpful. The series Moonhaven (Ocko 2022) depicts a colony sent to the moon with an artificial intelligence and mission to create a society that can heal the manifold problems upon the Earth. One of the primary things this society does is occulude blood relations and create foster relations in order to get over, or passed the violence/domination that is named, "blood begetting blood." More generally, the writer Sophie Lewis (2022) has shown a long radical tradition through Marx and Engels, former slaves in Southern United States, feminists, and queer radicals, who all saw the nuclear family as a source of oppression and false scarcity of care.

Trading Justice

Another strand of neoclassical economic literature models wealth and illth transfers through overlapping generations. These studies often involve trade-offs between present and future generations. A divide within these studies is whether to use a specific social discounting rate, where future generations are valued lower than present generations, which Caney (2014) compares to gender or ethnic discrimination. However, social discounting is broader than official discount rates and takes various forms. Caney (2014) groups social discounting into *pure time discounting*, which can be anything from 0.5 to 3%, and this can be fixed or variable; *growth discounting*, where perpetual economic growth is assumed and that future generations will inevitably be richer and so can pay for mitigation; and *opportunity cost discounting*, where it is assumed future technological advances will make it cheaper to delay climate action (Caney 2014). Thus, while some of the accounts detailed below reject the specific policy of a discount rate, many do include these other forms of discounting and so introduce justice trade-offs.

Some mainstream economists advocate for discount rate but with various caveats. Of note is the Stern Review's (2007) use of a 0.1 discount rate. Nordhaus (2007) argues that although this results in rapid CO^2 reductions, it leads to low rates of return on capital and high costs on a carbon tax. Instead, Nordhaus offers a rate of 1% per year in one model or a model calibrated to "ensure it produces the market return to capital," with his overall argument being that the Stern Review was too categorical and unambiguous and that policy-making needed to balance the benefits and costs of mitigation over time and not offer simplistic solutions (Nordhaus 2007, p. 699). Elsewhere Nordhaus (Nordhaus & Boyer 2000, p. 16) advocated a more specific reducing yearly discount rate of 3% in 1996, 2.3% in 2100, and 1.8% in 2200.

Beckerman and Hepburn (2007) reject the idea of zero or near-zero discount rates due to intergenerational justice based on two arguments. The first is what they label "agent-relative" ethics that, via Hume, argues we care for our closet family and then this care reduces further from this in relation or time. Interestingly, this perspective justifies the presenteeism and rentier empires that largely characterise current policy-making and our economy. The second is a form of the non-identity argument, in which they claim that as of now there is no agreement of what is just or proper, and there will inevitably be less agreement further into the future (Beckerman & Hepburn 2007). A more recent paper (Cisco & Gatto 2021) conducts an overlapping generations model focusing on climate justice using 0.99% yearly discount rate. Interestingly, they posit a steady state economy in order to achieve climate justice and intergenerational justice as an aspect of this, and argue that a business-as-usual approach or maintaining perpetual economic growth would require more work and less leisure for younger generations (Cisco & Gatto 2021).

Others opt for a variety of fiscal policies as fairer than social discounting. For instance, Sachs (2014, p. 249), rather than discounting, advocates deferring the cost by debt financing mitigation where "it may be possible to fund today's actions with public debt, so as to shift the ultimate costs of mitigation to later generations." He goes on to model various scenarios of this and then concludes that the current generation can debt finance mitigation and remain as well off but can act like a steward for the future but not sacrifice for it (Sachs 2014). In contrast, Howarth (1998), while also rejecting a discount rate, opts for rising carbon taxes and wealth transfers towards future generations. He models three approaches: a business-as-usual approach, a utilitarian optimum with the rising carbon tax and wealth transfers, and a suboptimal utilitarian option with rising carbon tax and no wealth transfers (suggesting political infeasibility). The second model creates the most fair and sustainable outcome with modest sacrifices in the near term and best outcomes on a range of climate and consumption factors. Howarth (1998, p. 589) concludes that this optimal path and policy "hinges on value judgements that are difficult to resolve within the framework of economic analysis."

Whether rejecting a discounting rate or not, these arguments rest on several assumptions. The first is that economic growth will and should continue (with one exception). This is a form of growth discounting whereby the assumption of perpetual growth justifies deferring the payment of mitigation until the future where we will all be richer. It is not hard to see the problems with this: firstly, there is strong evidence that economic growth is the problem as it cannot be separated from material throughput and thus planetary boundaries

(Vettese & Pendergrass 2022; Oliveira 2022; Hickel & Kallis 2019). Also, it violates the polluter pays principle (PPP), a well-established environmental norm (Broome 2012). Regarding the PPP, Davidson (2012) invokes "the reasonable person standard" in tort law to apply this principle to social discount rates and transgenerational environmental justice protections. He argues that is imperative to adhere to consistent norms in addressing pollution transcending generational boundaries, aligning with established international laws governing pollution across national borders. This contention underscores the necessity to employ a zero social rate of time preference in the cost–benefit analysis of climate policy. Such an approach advocates for non-discounting of climate damage incurred by future generations, refraining from adjustments based on anticipated wealth accumulation or temporal remoteness. In short, Davidson argues that there is no reasonable justification to apply different sets of standards to pollution experienced across generations as there is to pollution experienced across national borders. Within this analysis, from a legal and ethical perspective, social discounting violates tort law.

As a moral problem, of note is the language that surrounds economic valuation of future lives and the inherent imagination and conceptualisation of future people. The language of consumers, utility and utility maximisation, smuggles in the figure of *Homo economicus* – a much-critiqued caricature of a more complex human with multiple desires and motivations beyond utility maximisation (Hickel 2020; Lennon et al. 2019; Graeber 2001). Finally, these economist accounts of intergenerational justice treat humanity as a set of homogeneous "publics," consumers, or voters without differentiation of responsibility of nation, class, gender, ethnicity, ableness, or per-capita emissions. This homogeneity in the economic and political imagination occludes the normative argument that the impoverished, be that within the minority world or in the majority world more generally, should be able to push the costs onto the wealthy (Caney 2014), who are far more responsible for the CEC (Hickel 2020; Chancel & Piketty 2015). These issues with mainstream economists thinking on intergenerational justice are part of a broader problem with economic thinking surrounding the CEC, to which we turn next.

The Economics of the Anthropocene

Kim Stanley Robinson in *The Ministry for the Future* (2020, p. 106) argues:

> The whole field and discipline of economics, by which we plan and justify what we do as a society, is simply riddled with absences, contradictions, logical flaws, and most important of all, false axioms and false goals.

Robinson is clearly referring to mainstream or *neoclassical* economics. This section aims to show why in many ways this reading of neoclassical economics is right and that it is a dangerous discourse when it is applied to the CEC.

The first part will deal with the overall myopic epistemology of mainstream economics and how this attenuated over the 20th century (Kessler 2015). This attenuation was paradoxical as the ideas and methods of economics expanded and colonised many other fields of enquiry (Sahlins 2017). The second part will show how this blinkered epistemology impacts the empirical work of mainstream economics on the CEC and how this produces quite dangerous complacency that will endanger lives. This would not matter so much if nobody listened to these people or this discourse. However, as Robinson's quote above implies, economics is a kind of master discourse that has the ear of power while also insidiously colonising other fields. This suggests we need to listen to what is said and oppose ideas when they are so clearly dangerous and damaging.

Economics as a field of enquiry and the economy as an object of concern and study have had a relatively short career. Mitchell (2013) shows how the idea of a national economy emerges at the beginning of the 20th century and really takes hold during the interwar years as the big empires of Europe begin to fall apart. Before this in the British empire "where the sun never sets," it made little sense to frame this economy as discretely national as it was extracting resources and exploiting people across the world (Malm 2016b). The field of economics emerges from the older field of political economy and starts to be applied to this new concept of a national economy. This earlier idea of political economy was the overall governance of a polity, while in contrast economics and the economy become discursively divorced and isolated from its wider conditions of possibility within a polity. A good example of this is the mainstream ideas of negative externalities – the pollution or negative social impacts that occur from economic processes. The term externality implicitly separates the economic and economy discursively from the wider world.

In its infancy, economics had a plurality of epistemological approaches, but relevant here were two broad strands. On the one side (Mitchell 2013), there were those who largely followed in the tradition of Jevons, the 19th-century British political economist. These thinkers start from resources, like coal or wood, and tie these to the human and social activities these resources support and allow, as well as the limitations they present through overuse. On the other side, building on the statistical work of Jevons, there was British economist Keynes who moved the focus from natural resources to the role of money and flows of capital. Also, in the United States, there was eugenics champion Irving Fisher, who argued that it was not overuse of resources but the degeneration of racial purity that was the problem, generating a string of unjust economic policies in the areas of immigration and labour reform in his wake (Leonard 2009). The latter side won the debate, shifting economic knowledge from the study of resources and human use of these, to the study of prices, flows of money, and the behaviour of very narrowly defined humans (Mitchell 2013).

Another aspect of economic epistemology concerns how it deals with uncertainty. Kessler (2015) shows how Keynes distinguished between uncertainty and risk, whereby risk involved known probabilities such as in a game of roulette, while uncertainty was the realm of the human and our collective future to which there was no way to mathematically model this. This begins to change in the 1950s and 1960s with the work of various economists who argued all this uncertainty could be probabilistically quantified in various mathematical calculations. This results in a methodological drift, memorably labelled creeping "off into the thickets of algebra" by Joan Robinson (2006, p. 110). This involved the profession taking in more mathematicians and physicists able to do this algebra, and an institutional lock-in of these methods via publications, degrees, and careers (Kessler 2015).

A final point to make here is the performative nature of this field of economics and how this affects and shapes its object of study, the economy, and the people within it. MacKensie's (2006) study of financial economics shows how this performativity works. For instance, the efficient-market hypothesis suggested mature capital markets instantaneously incorporate all relevant information into and through prices, thus implying individual financial bets on specific stocks were largely pointless. This hypothesis' early results did not reflect the market, but this increased in accuracy over time, without any change to the underlying theory. What had changed were many individual financiers who had begun to match their investments to the efficient-market hypothesis. Emerging from this hypothesis was the index fund that tracks the whole stock market and its gradual growth (Mackensie 2006; Raworth 2022). The power of this performativity is seen in the massive growth of these "passive" index funds and in how such financial behemoths as BlackRock, Vanguard, and State Street (the "Big Three") routinely invest in these index funds. An important consequence of this is "Big Three" investors were estimated in 2017 to be the largest shareholders in 88% of S&P 500 companies. This gives a very small number of powerful financiers voting rights and influence over vast areas of our lives and world (Fichtner et al. 2017; Bebchuk & Hirst 2019). Overall, economics over the 20th century narrowed its epistemology to the study of prices, flows of capital, and the utility maximiser, while methodologically moving into the thickets of algebra. Simultaneously, as the field gains prestige and power, it begins to shape the world it ostensibly describes with important impacts on the economy, the world, and human subjectivities.

The irony behind all this is the economics and economy that emerge from WW2 that occlude its own materiality and, in fact, become even more reliant upon the object of oil. It is oil that underpins the great acceleration of the post-WW2 era and the associated cumulative emissions (Mitchell 2013). This ironic occlusion of material dependence while deepening this dependence is perhaps at the root of what has been dubbed the imperial mode of being or the fossil fuel subjectivity (Roos 2023; Eversberg & Schmelzer 2023). This is a mode of life or subject that sees themselves as individual and independent from the manifold interdependencies of modernity. This works on many levels, from denying our fundamental human vulnerability and thus reliance on social relations of care, to denying our and much of the wider web of life's reliance upon the ecological niche that capitalism, economic growth, and its climate impacts are destroying. This fossil fuel subjectivity can be seen in mainstream economics today in the discourse of both ecological modernisation and prominent economist who have focused on the CEC.

The Ecomodernist Manifesto (Asafu-Adjaye et al. 2015) leans into the idea that we can be independent of "nature," going even further by arguing we need to leave "nature" to save it. This leaving involves for ecomodernists the intensification of "farming, energy extraction, forestry, and settlement" to reduce land use and environmental impacts (Asafu-Adjaye et al. 2015, p. 7). To do this, they employ the notion of weak sustainability, whereby anything is fungible and replaceable via human ingenuity. Thus, they claim the notion of "fixed physical boundaries to human consumption to be so theoretical to be functionally irrelevant . . . Given plentiful land and unlimited energy, substitutes for other material inputs to human well-being can easily be found if those inputs become scarce or expensive" (Asafu-Adjaye et al. 2015, p. 10). This manifesto frames the CEC as essentially a technological challenge, thus offering two techno-fixes that will lead to a "good Anthropocene." Firstly, it claims the sun offers more than enough energy that could be harvested by more efficient solar cells and battery storage. Secondly, nuclear fission is claimed to be "zero carbon" and has the required energy density to provide the intensification this vision demands – however, because of "social, economic and institutional challenges," a new generation of "safer and cheaper" nuclear technologies will be required (Asafu-Adjaye et al. 2015, pp. 17–22).

The problems with these ideas and the discourse of ecomodernism are Robinson's "absences, contradictions, logical flaws" and of course, false axioms and goals. In terms of absences there is no understanding, or an elision, of the materiality of technology and how in the past and present this was integral to shifting costs and illth across space and time (Harris 2023; Roos 2023; Parrique et al. 2019; Hornborg 2009). This can be seen in both technologies chosen by the ecomodernists, with solar PV goods flowing North to the minority world, while the wastes and mining flow South to the majority world (Roos 2023; Sovacool 2019). Similarly, Hecht (2012) shows how the damaging and dangerous mining of uranium done in the majority world is discursively organised out of the nuclear applications it allows in the minority world. More specifically, Voyles and others (2015; Powell 2018; Gómez 2022) detail how the Navajo community and the region of New Mexico have been "wastelanded" by uranium mining and weapons testing. There is also an implicit push of today's problems into the future with the call for better Solar tech and the classic silver bullet, nuclear fusion.

In terms of contradictions, the manifesto treats "nature" as some external objective fact. As an external object, we can separate ourselves from this "nature," which essentialises humans as in some way beyond, independent of, and above nature. Also, the idea that we can intensify urbanisation and industrial agriculture without this impacting "nature" while we remain separate from nature is contradictory, similar to the implication technologies descend like mana from the sky. In connection, the lack of mention of the alarming loss of biodiversity, linked to industrial agriculture and urbanisation, makes one question what the "eco" of this manifesto refers to (Crist 2016). Similarly, the manifesto talks of the need for growth and development to alleviate "poverty," another category that is presented as some unrelated, essentialised thing, rather than a consequence of the social relations of neoliberal capitalism (Mau 2023). If poverty could be alleviated by more of the same, why has this not happened? The country to make the most significant progress in the last four decades on reducing poverty is China (World Bank 2020), and this is not, at least officially, a capitalist country. Finally, the manifesto implies that the global poor and indigenous groups are as destructive to the environment as multinational corporations, which is simply not accurate. This is not to suggest that subaltern groups do not harm the environment but that this is incomparable to multinational mining, fossil fuel, and other corporate impacts. Also, there is substantial evidence that native peoples, and impoverished people more generally, tend to form the core of the opposition to corporate and state environmental degradation (Harris 2023; Frank 2022; Estes 2019; Bonneuil & Fressoz 2016). Thus, this manifesto by this peddling and this obfuscation contributes to a longer racist and Malthusian environmental strain going back through eugenicist Garrett Hardin (1968) and various elite White US conservationists such as Theodore Roosevelt (Cronon 1996). This strain blames environmental degradation, species loss, and the CEC on the global poor, the racialised, and the Other, placing environmental problems more generally at the "border" and concerning of "undesirable" immigration and population (Malm & The Zetkin Collective 2021).

Lastly, in terms of false axioms and goals, we might question the idea of the "good Anthropocene." Good for whom is an initial question. Clearly, it will not be good for the more-than-human as this notion leans into the essentialisation of the human as the master of this world and all flora, fauna, and "ecosystem services." It also speaks of a hubristic Promethean mentality that misunderstands our place in this world. As the climate system destabilases as this is written, who really thinks we are in control of ecosystems and the biosphere? We barely understand them, much less control them (Vettese & Pendergrass 2022). Rather in the Anthropocene, we should be cognisant of the active part of what we call nature will be, and how this is not a linear process we can be complacent about, but one involving potential "catastrophic state shifts" (Dryzek & Pickering 2019, p. 10). Finally, ecomodernism is an attempt to save capitalism from itself, and thus the "good Anthropocene" is a continuation, with tweaking, of a system that even without the CEC is benefiting a tiny minority of the global population and so lacks legitimacy anyway

(Harris 2023; Táíwò 2022; Jackson 2017). However, as bad as ecomodernism is in terms of justice, certain neoclassical economists working on the climate crisis are even worse, and so will be explored in the final part of this economic critique.

We have not yet surpassed 1.5 degrees centigrade average increase in the global temperature, yet wildfires, flooding, and severe storms are already causing dramatic loss of life and livelihoods across the world. Nevertheless, prominent neoclassical economists try to suggest higher average temperature rises would be *optimal* in their endeavours to "balance" economic priorities against climate priorities. However, the problem here is not just the output of these economists but again the methods and epistemology employed. This part will focus on Keen's (2020) critique of climate-focused neoclassical economists, for it focuses on both questionable processes and outcomes of this research.

A first problem Keen (2020, p. 2) identifies is the highly questionable use of methods and data, which is described as "making up numbers to support a pre-existing belief." Namely, this belief is that the CEC will have an insignificant effect on the US economy. The first questionable method is *enumeration*, which involves putting a number on the cost or damage to an economy from the CEC. This number or cost is obtained from "climate models, impact models and laboratory experiments" (Tol 2009, pp. 31–32). Yet this starts from assuming that the CEC effect will be trivial, with Nordhaus (1991, p. 930) claiming that "it must be recognised that human societies thrive in a wide variety of climatic zones" and that other determinants such as labour skills of technology "swamp climatic considerations." Nordhaus (1991, p. 930) goes on to then exclude 87% of the US industry, such as finance or government services, from climate impacts simply because it "carefully controlled environments." Arguing against these points is not difficult – what applies in a stable climate does not necessarily or likely apply in an unstable one. Or, arguing that 87% of the economy goes on in controlled environments ignores the fact that many activities, such as farming or fishing, while not highly valorised under capitalism remain essential for social functioning. One might ask what relevance a 1991 paper has here, but this analysis and the underlying assumptions were almost verbatim reproduced in the 2014 IPPC supplementary report (Arent et al. 2014, p. 662). Moreover, Nordhaus won the "ersatz" (Mirowski 2020) Nobel Prize for economics in 2018 saying very similar and dangerous things (Hickel 2018; Keen 2020).

Another questionable method is labelled by Tol (2009) as the statistical method, but Keen (2020) describes this more accurately as a cross-sectional approach. This in essence compares how per-capita average income varies across space with very different average temperatures – say Florida versus Alaska, and then uses this to estimate the impacts of future climate warming (Tol 2009; Nordhaus 2006). Assuming climate differences over space can be used to model climate warming over time, firstly, contradicts the enumeration method, which discounts climate as a factor as outlined earlier. Moreover, this method creates the idea of an optimum economic productivity temperature, which the CEC will move some places towards, when they move upwards to this temperature, and some places away from when they go beyond this temperature (Keen 2020). Emerging from this is the idea of the benefits of the CEC and endeavours to quantify these benefits (Mendelsohn et al. 2000). These benefits mainly accrue to the minority world and are thus an incentive to inaction. Or, as Keen (2020, p. 6) argues, the initial assumption – climate will have little effect on the economy as long as we do not spend too much now on mitigation – becomes your conclusion – that we should pick the low-hanging fruit and then leave mitigation to the future.

A final method used by Nordhaus and colleagues was the use of expert opinion, but again used in highly problematic ways. Keen (2020) shows how this was used in a study by Lenton et al. (2008) that gives specific guidelines on using this method. Importantly, in Lenton's study, which included 52 respondents, the experts were told to stick to their area of expertise. However, in Nordhaus' (1994) study, of the 18 people included, 10 were economists, with eight of these not ecological or environmental economists. Thus, according to Lenton et al. (2008), eight of 18 people should not have been included in this study. After all, what is the point of asking an expert for opinion beyond their expertise? Surely, this just becomes lay opinion.

There are a number of deeply problematic assumptions guiding this kind of research and discourse. Broadly, DeCanio (2003) argues that the CEC is a global crisis that risks disrupting Earth systems and that the general approach to cast it in terms of economics is surprising and a mistake. This is because economics and its models are good at predicting and measuring marginal changes, not systemic change (DeCanio 2003). Thus, it is not surprising that mainstream economists tend to argue for tweaking and incremental change. However, there are more problematic assumptions involved here; the first being that these mainstream economists claim they are value-neutral. This employs Haraway's (1988) god trick of claiming to be outside the system one describes, even while mainstream economic theories have profound performative effects on our world. Then there is a highly racialised assumption that lives in the majority world are expendable when they likely become unliveable under these *optimum* temperature rises. Similar is the anthropocentric assumption that biodiversity and the world's flora and fauna that have not been valorised and quantified are also expendable for it can be replaced.

Finally, we have the growth fetish and the attendant growth discounting that assume growth will and should continue. This assumes everyone will be richer in the future, and thus they can pay for climate mitigation and will likely also have better technology to do so (Caney 2014). Again, this assumption is not hard to question as in both the United States and the United Kingdom, this growth has not led to all people getting richer. In the United States in 1975 for instance, the GDP per capita was half of the 2018 level, yet on average people had higher wages and were generally happier, and the poverty rate was lower (Hickel 2018). In terms of wages in the United Kingdom, there is a similar dynamic with the economy growing, but in real terms average wages have been stagnating since the 1970s and there have been associated rises in levels of poverty, especially after the imposition of austerity after 2008 (Dorling 2023; Jackson 2017; Soper 2020; Davies et al. 2022). This upward extractive dynamic also appears to apply to the wider world. This occurs through the last 50 years but is then exacerbated after the 2008 financial crisis and then again during the pandemic, producing an ever-widening gulf between the very wealthy and the rest (Davies et al. 2022; Oxfam 2022; Parrique et al. 2019; Jackson 2017). Another problem with economic growth is its neocolonial elements. A significant driver of the minority world's growth since formal decolonisation beginning in the 1950/60s can be seen as neo-imperialist unequal exchange at the cost of the majority world. This involves the unequal transfer of land, labour, energy, and raw materials from the majority to minority world amounting to approximately $62 trillion over the period 1960–2017 (Hickel et al. 2021; Hickel et al. 2022; Hornborg 2009). These resources could have been used to deal with extreme impoverishment that affects many places in the majority world, which emphasises the injustice of this transfer of wealth (Hickel et al. 2021).

Economic growth, even if it was not directly driving the CEC, which it is (Jackson 2017; Hickel & Kallis 2019; Keen 2020; Stoddard et al. 2023), would be one of Robinson's false social goals. This is because beyond a certain point where people's needs are fully met, it is unnecessary, or people do not report increases in well-being (Inglehart et al. 2008). Also, there are other less destructive paths to human fulfilment via care, community, and senses of autonomy and purpose (Dorling 2023; Soper 2020; Dryzek & Pickering 2019; Jackson 2017; Kimmerer 2013). However, in many places this modest achievement of meeting people's needs has not been met. Simultaneously, vast amounts of wealth and resources are hoarded by the extremely wealthy and used to undermine democracy and climate mitigation (Dorling 2023; Mirowski 2013; Aronoff 2021; Kenner 2019). This does not mean the majority world does not need to develop socially useful and essential infrastructure or that in the minority world we do not need more care and renewable infrastructure. What it does mean is we need to reduce and phase out harmful things like fossil fuels, plastics, and high-emission activities (Buck 2021), while redistributing resources on ultimately a finite planet (Táíwò 2022; Jackson 2017).

Finally, it is crucial to examine what kinds of *optimum* average temperature are being presented by these mainstream economists. In 2018, in the same year he won the Nobel Prize for economics, Nordhaus (2018a) ran and compared his DICE integrated assessment model (IAM) against seven other IAMs, which produced an average of 3.87 degrees centigrade warming by 2100. In his presentation accepting the Nobel Prize, Nordhaus (2018b) suggests that an optimum global average warming should be 4 degrees centigrade. More concerning is a debate reported and reproduced from social media in which Richard Tol defended the cross-sectional approach arguing that "climate determinism" was empirically unfounded. In this debate, Tol seems to defend a 10-degree-centigrade average warming, saying it would be manageable by moving indoors "much like the Saudis have" (Keen 2020, pp. 16–17). As argued earlier, we are not yet past 1.5 degrees average warming, and we are already seeing tragic impacts of the CEC unfold. Also, we may be nearing tipping points, or non-linear exacerbations of the CEC, non-linear processes that this economic modelling systematically fails to incorporate (Keen 2020).

Overall, what we see here are very dangerous conclusions made from problematic data and methods – or as Keen (2020) says, this is simply a case of "garbage in garbage out." Moreover, it is based on founding assumptions and a worldview emerging from powerful people in the minority world that discount and devalue future generations, the majority of the world's population, and the wider web of life. Thus, to return to Robinson's quote above, we might say the neoclassical economics evidences the problems he claims, but rather than it produces dangerous axioms and goals.

Conclusion

This chapter has examined how mainstream economic thinking is a problem in justice terms as it pits today's generations against each other and against future generations. It also either ignores or at best instrumentalises a more-than-human nature, thus dangerously downplaying the CEC. Epistemologically, it is infected by Bhaskar's (1997) epistemic fallacy – mistaking its models of reality for our social reality. This would not be such a problem if this did not have such damaging performative effects on the economy at large, and also at the individual subjectivity level. As such, economics can be seen as the empirical science of liberal philosophy, moving from Descartes *I think therefore I am* to *I measure therefore it is thus*. This philosophy and mainstream economics are failing us, so we need new forms of thinking, doing, and being. The practical aspects of this will be the subject of the next chapter, where we attempt to outline, via a range of practices, a more substantive and inclusive form of justice.

4 Abundant Justice and Democracy

This chapter will explore more novel thinking and approaches to intergenerational justice. The argument will parallel that of the former chapters following the simple and logical presumption that the techniques and thinking that get us into a problem are perhaps not the best for getting us out of the problem. This idea can be applied to the problem of the CEC, largely caused by markets and endless growth, which we are told will be solved by "green growth" and carbon markets (Lohmann 2013). Or, as we saw in the last chapter, intergenerational justice can be solved by loading debt and costs onto either today's children or future generations. However, doing so largely ignores the current debt mountain as one aspect of what might be called the externalising machine of capitalism, piling up debts for the many, just as it does physical waste. Regardless, as argued throughout this book, justice for one group at the expense of another is not justice but extraction.

Therefore, moving from the old to the new, this chapter will document quantitative researchers working around the old prisoners' dilemma reworked as an intergenerational dilemma. This research suggests talking about future generations helps build intergenerational solidarity with some interesting techniques used to stimulate this thinking and solidarity. It will then move on to research that considers children and today's youth as proxies for unborn generations. Theoretically, Davies (2012) has conceived of intergenerational democracy, from which she and colleagues have conducted empirical work on how this might be realised in a world that discounts both the very old and the young. We will then examine some empirical studies of children's action on climate. This includes how the mainstream media have treated the Friday's for Future (FfF) movement and how children rights have been addressed internationally at the UN. We will then collate the method of intergenerational democracy with ideas from Chapter two and these empirical studies to present a substantive and inclusive form of justice through an adapted version of the capability approach. The final section will outline a method of deliberative backcasting future study that will fit this substantive and inclusive model of justice and perhaps help us move towards this.

DOI: 10.4324/9781003422686-6

Intergenerational Dilemmas

A reworking of the prisoner's dilemma employs and delivers more promising techniques and findings for intergenerational justice. In these studies, they employ a game where in rounds multiple or single players are a particular generation in a chain and must choose between investing or not investing resources into sustainability. During this process, participants generally have to conduct social value orientation activities to be characterised into socio-demographic, criticality, and empathy groups for cross comparison. Traditionally, a key factor in the prisoner's dilemma was whether the prisoners could communicate as part of their decision and thus act in concert and cooperate (Kagel 2018). This is impossible or only one-way between non-overlapping generations and a major dilemma of intergenerational justice. It is how they addressed this dilemma that is the most interesting innovation within these studies. Koirala et al. (2021) conducted an intergenerational sustainability dilemma experiment with Japanese university students. As their treatment they had three test groups: majority voting where each participant in a single generation vote A or B with totals aggregated; deliberative majority voting (DMV) with the same process above but with 10 minutes of deliberation before voting; and majority voting with deliberative accountability (MVDA) where they do both majority voting and DMV and produce a collective written reason for their generation's possible decision and advice for future generations. Via descriptive and inferential quantitative analysis, the authors show that the DMV and the MVDA treatments increase participants choosing the sustainable intergenerational option. Descriptive statistics shows an increase in participants taking the sustainable choice from 5.71% in majority voting to 12.12% in DMV and to 27.78% in MVDA. A logistic model shows MVDA groups are 5% more likely to choose the sustainable option than DMV (Koirala et al. 2021).

A contrasting study which chooses a one-person intergenerational sustainability dilemma game and a future ahead and back (FAB) treatment is also instructive (Shahen et al. 2021). This experiment follows a similar process to the above, but this time there is no group communication. Nevertheless, the FAB treatment is also innovative in the way it endeavours to overcome the fact that non-overlapping generations cannot communicate. It does this by asking the FAB treatment group to imagine they are the next generation and to request what they want from the previous generation, after which they go back to the original generation and make their decision. The results follow a similar pattern to Koirala et al. (2021), with the FAB individuals choosing the intergenerational sustainability option 44.5% of the time compared to 33.7% in the basic game. Shahen et al. (2021) conclude that the FAB effect could be due to cognitive dissonance triggered by playing both roles, or an increase in non-cognitive aspects of decision-making such as empathy.

Both these studies offer innovative ways of activating intergenerational empathy and solidarity while also evidencing the power of deliberation and public participation more generally. Cotton (2017), in the context of the United Kingdom's long failure to deal with its nuclear waste, argues that real progress was made only in the 2000s when the then Labour government introduced extensive and unprecedented (if not ideal) deliberative processes to deal with nuclear waste (Cotton 2017; Chilvers & Burgess 2008). The so-called *deliberative turn* discussed throughout this book represents a concerted effort to remedy the invisible injustices hidden within the classical economic conceptualisation of benefits and harm. Allowing opportunities for people to come together through a process of dialogue and collaboration gives recognition to the notion that non-economic values, including ethical, aesthetic, and cultural values, should have equal weight within environmental governance across multigenerational time frames (Ercan and Dryzek 2015). More generally, deliberative democrats have long argued that progress can more readily be made on complex and intractable problems using these methods. Dryzek (2000) argues that we need to deepen our attenuated democracies in three interacting areas: making as many participate as possible with particular focus on those commonly disenfranchised; widening the scope of issues we democratically decide upon such as voting on budgets or taxation; and finally, increasing the authenticity of these exercises for it is deeply damaging of democratic trust when a deliberative democratic decision is then ignored and belittled by elected officials (Geels 2014; Dryzek 2000).

In describing what he calls *ecological democracy*, Dryzek (2013) elaborates on how local deliberative democratic methods can improve socio-environmental decision-making in practical and justice sense:

- They effectively integrate differing perspectives and values on complex issues.
- They help generate and collate feedback mechanisms within socioecological systems.
- They sustain interests of communities over selfish interests within or without this community.
- They allow concern and empathy for those not there or present, such as future generations or non-human entities.
- They help generate creative solutions by addressing core concerns of conflicting positions.
- They help connect general theories and principles to specific examples and cases; finally, they may help self-reflexivity where participants reflect upon their own norms and positions and re-evaluate these and perhaps become more environmentally concerned (Dryzek 2013; Dryzek & Stephenson 2011).

Thus, deliberative democratic methods offer much, but to achieve substantive justice, these methods need to be reflexive in process and in who is involved and how people are involved. The next section will expand on these aspects.

Children and Young People as Future Generational Proxies

Democratic exclusion of the young has a long history with ancient Greece, seen in Eurocentric discourse as the cradle of democracy, excluding everyone from the polis but free males aged over 18. Also, important here is that during the "democratic" periods of ancient Greece, representative democracy was seen as elitist and subject to manipulation by wealth and power, while sortition, or selection of citizens by random lottery for office, was preferred as less corruptible (Graeber & Wengrow 2021; Taylor 2019). Both issues of exclusion of the young and the problems inherent in representative democracy are addressed by current practitioners of deliberative democracy, who argue three elements of deliberative democracy can alleviate these problems. These three elements are a non-random democratic lottery to select participants who are representative of communities including those generally excluded, for instance young people; good conditions for deliberation, including appropriate time, payments for time, and respect for all; and final recommendations being formed by super majorities of over 70% with the aim for transformative solutions based on common ground rather than trade-offs (Cesnulaityte et al. 2022).

This section will see how theorists and practitioners have worked to include children and young people in dealing with the CEC. This is an international endeavour just as the CEC is an international set of interlinking crises, with the nation-statism of our world being part of the problem. The first section will explore the work of Kirsten Davies on intergenerational democracy, a perspective that recognises that current forms of representative democracy discount the young and the old. This can be seen to parallel the myopia of presenteeism, sure of its own perspective and needing neither the old/history nor youth/future to aid its bounded rationality. The second part will cover how children and the young shape and are shaped by media representations, with FfF a key social movement active since 2019, arguing forcefully for more youth involvement in crucial decisions that will impact their lives. The next part will explore a comparison of activist discourses emerging from two young groups – FfF, a movement distributed around the world, and the UN's Major Group for Children and Youth. The final part of this section will look at how children have been dealt with by the UN Convention on the Rights of the Child (UNCRC).

Intergenerational Democracy

Intergenerational democracy (Davies 2007) takes a broad and holistic view of human and more-than-human relations across time and space. It recognises that democracy, capitalism, and individualist societies are root causes of the CEC (Davies 2007) but argues that a localised, more direct form of democracy can be part of the solution. In this sense she parallels, if not going as far as, decolonialist thinkers such as Mignolo (2007; Escobar 2010), who argue that the package deal of modernity, democracy, and progress is built on but occlude the colony, prison, and gendered, ablest, and racialised oppression, and that the way out of this is via a localised pluriverse of new liberational ontologies. As an "all inclusive" approach, intergenerational democracy focuses on lifecycle with an aim to "assist the reconstruction of generational and environmental relationships by engaging whole communities, from children to the elderly, in planning and managing their futures" (Davies 2007, p. 71). Arguing that the three pillars of sustainability, economic, environmental, and social, largely ignore the local culture, intergenerational democracy aims to democratically engage those members of communities that are mis- or unrecognised in policy and planning forums (Davies 2007; Davies et al. 2016). As such, intergenerational democracy works through the three themes of environmental/energy justice (distribution, process, and recognition), specifically tying intragenerational and intergenerational justice together.

Work on intergenerational democracy began in Australia and then continued in Vanuatu with methods and findings from these studies relevant to this book. Methodologically, intergenerational democracy follows three broad stages, with the first being gaining funding and support from the local authority and forming an intergenerational sample group. Following this is the consultation and planning stage, which can be quantitative or qualitative depending on project aims but crucially must not randomly select participants but purposively select to reflect the community demographic (Davies 2007). In this way, intergenerational democracy follows the latest innovations in deliberative democracy across the world, which as stated earlier involves as its first principle this non-random form of sortition (Cesnulaityte et al. 2022). During this consultation and planning stage, surveys are used including written/visual responses; for instance, primary school children are invited to submit visual portrayal of the issues. Subsequently or contemporaneously, interviews are used to develop support and community ties, but are then followed up with the more social methods of focus groups and larger community forums (Davies 2007; Davies et al. 2016). The final stage is the implementation stage, with this being the development of a 25-year plan for the local area in the first project (Davies 2007), which involves ongoing involvement of representatives of the intergenerational cohort. Again, intergenerational democracy corresponds here to both the latest in deliberative democratic methods and processes and Dryzek's third theme of deliberative democracy – to make the process authentic and transformative (Cesnulaityte et al. 2022; Dryzek 2000, 2013).

In terms of findings, intergenerational democracy finds many differences and similarities across generational cohorts, similar to an intergenerational justice project conducted by one of this book's authors (Ward et al. 2021).

In the first case study based in a suburb of Sydney, notable differences were younger groups tending to show more concern for the CEC and expressing concern that they were excluded from policy-making that would directly impact their futures. Thus, one 10-year-old argued (Davies 2007, p. 82):

> We should work to a level where children's views are regarded just as important as any adults as we are the ones that shall be living the future.

Similarly, in a second case study in the central west of New South Wales, it was the youngest groups who expressed the most concern for the CEC. An older youth group (11- to 20-year-olds) when asked about their concern about the impacts of the CEC reported 26% as neutral; however, when this was followed up in interviews, this was largely reinterpreted as an anxiety avoidance strategy induced by the threat of the CEC (Davies et al. 2016). Anxiety and displacement are common and predictable responses to an imminent threat. However, rather than accepting this as a predictable social response and a more general social anxiety, there is a damaging tendency to pathologise individuals (Ojala et al. 2021). Older groups in both Australian studies showed concern about the CEC, but to a lesser degree, while being more concerned about other issues. One notable issue was immigration and maintaining the heritage of the local area (Davies 2007). This is ironic for White settlers arguing to maintain a conservative, Christian culture on colonised land and is a specific example of forgetting and presenteeism in a modern representative democracy. A final contrast in the first Sydney case was gender-based differences. These differences perhaps reflect the way our societies impose gender normativity upon children from a young age, as feminists have long argued (Lewis 2022; Fraser 2014; Mies & Shiva 2014). In a rather classical formulation, Davies (2007, p. 79) claims "[F]emales were focused on care-based initiatives, while males outlined the importance of governance and technology."

A further important case study of intergenerational democracy was conducted on the island of Espiritu Santo, Vanuatu. Davies (2015) follows a broadly similar approach to that detailed earlier, but this case involves subjects more threatened by the CEC, specifically rising sea levels and pollution. Customary law within Vanuatu appears to value intragenerational, intergenerational, and more-than-human aspects of justice. This is evident in Vanuatu's constitution which echoes the Brundtland Commission to "protect the Republic of Vanuatu and to safeguard the national wealth, resources and environment in the interests of the present generation and of future generations" (Davies 2015). In contrast to the Australian case, there was much more shared concern for the CEC and its impacts. This cross-generational concern might not only be because these island nations are existentially threatened by the CEC but also because although this nation exists partly in the modern world, it also retains elements of customary and indigenous ways of life. This was perhaps best expressed by one young participant who argued people must look after (Davies 2015, p. 63)

> [T]heir physical environment and making the most out of it for themselves as well as remembering to leave something intact for their future generations . . . Also learning to adapt intelligently to the modern changes that seem to be overtaking them in rapid succession.

This statement and broader studies show that it is not a question of rejecting all of modernity and certain benefits it brings, but one of recognising the unjust trade-offs and occlusions of the modern world. Overall, a strength of this intergenerational democracy approach is a focus on the damaging ageism of our societies that routinely discount the views and ultimately the lives of the younger and older groups. This book argues that any form of intergenerational justice must start with engaging these excluded groups and that the younger groups can be seen as proxies for future generations, an idea that will be elaborated later.

Media Framings of Youth Protestors

In 2019, young people across the world inspired by Greta Thunberg's example began to strike from school on Fridays against the lack of action or even maladaptation to the CEC. This became known as Fridays for future (FfF), and a number of studies have tried to understand this movement from various angles. These include media analysis, discursive comparisons between FfF and UN children forums, and an analysis of the UN communications procedure that is part of the UNCRC. A theme running through these studies is the ambiguous position of young people and children in the CEC, with all agreeing these groups will inevitably be more affected by this crisis, but the role these groups should play now, and in the near future, still being the point of discussion.

The modern corporate media and in many ways institutions like the BBC are not so much reporting news but, as Herman and Chomsky (2002) claimed, manufacturing consent. Zabern and Tulloch (2020), who explore the representation of young climate activists, list how this is done (largely following Herman and Chomsky's earlier insights). This is known as framing and includes

- selecting specific issues and presenting these in certain ways;
- selecting certain aspects of a topic and highlighting this with problem, cause, moral evaluation, and solutions offered (this often includes right-wing-funded think-tank actors who are rarely identified as such);
- working with political/economic elites who are routinely given platforms to shape a message that favours the status quo;

- and an overall creative process that does not just reflect the world but is a form of Haraway (1988) or Habermas' (1996) discovery and creation of the world, but in this case to the benefit of existing power structures (Zabern & Tulloch 2020; Herman & Chomsky 2002).

Thus, young activists tend to get misrecognised and critiqued in the mainstream media as these young people, by their activism and demand for agency, are challenging the status quo. Exploring German media representations of young climate activists, two articles use differing approaches but find similar results. Zabern and Tulloch (2020) use a framing theory approach – basically that outlined earlier, which is largely natural and the way we communicate, but when used by the corporate media becomes a positive feedback mechanism of power. This study looks at three German newspapers and how they represent the youth protestors and their demands. Zabern and Tulloch (2020, p. 12) document who gets to speak, with firstly the protestors (58 of 147 sources), followed by politicians (35), experts (17), educational officials (11), and finally parents (9). They find two groups of framing, with one fostering intergenerational justice, including David versus Goliath and climate threat frames. The group of frames undermining intergenerational justice and the protestors include truancy, activist without activism, or vice versa, whereby the aims of the protest are removed/depoliticised. Finally, there is Hamelin, a performative frame focusing on the climate celebrity, Greta Thunberg, and her "apolitical, truant fans," which again serves to trivialise and depoliticise the protests. Overall, the negative coverage outweighs the positive 50 articles to 21, with the truancy frame being the most common at 17 pieces (Zabern & Tulloch 2020, p. 20).

In contrast, Bergmann and Ossenwaarde (2020) employ a critical discourse analysis (CDA) method that collapses the binary of text/context and explores the crucial role discourse plays in reproducing injustice and domination in the world. Similar to the above, Bergmann and Ossenwaarde (2020) find the preponderance of coverage to be negative and disparaging and four connected thematic representations of the youth activists and Thunberg herself. These are the young heroine, the pupils, the dreamers, and the absentees. Firstly, the individualising theme of the young heroine involves a blame game for she is the leader of the absentees, but more generally there is an air of youthful naivete attached to this construction. The other three themes interconnect and support each other. The first, the pupil implying a lack of education and unrealistic claims, is similar to the dreamer who has nice ideas but needs to wake up. This is rounded off with the law-breaking absentee synonymous with the frame of truancy, suggesting unreliability and illegitimacy (Bergmann & Ossenwaarde 2020).

Other media studies add to or contrast with these German media studies. For instance, Jacobsson (2021) also employs a CDA approach in the Swedish media context, finding that the so-called objective nature of journalism removes conflict and social critique through a process that empties youth protest of its political substance, instead fusing this emaciated discourse into consensus (Jacobsson 2021). In this sense, journalists are agents of both the broader neoliberal operation of depoliticising the CEC (Methmann et al. 2013) and manufacturing consent/consensus (Herman & Chomsky 2002). In contrast, a UK study (Hayes & O'Neill 2021) of visual representations of youth climate strikers finds a move from an earlier representation of the climate movement as protest, which obscured the human face, to one that is more human, young, and female. They conclude that this depicts the CEC as questions of equity and intergenerational justice (Hayes & O'Neill 2021).

The role of the media in just and democratic societies has long been a concern of democratic theorists. Herman and Chomsky's (2002) argument that the media's role is one of manufacturing consent contrasts sharply to Sen's (2009; Habermas 1996) theoretical bastion of democracy, in which the media holds power to account and educates both elites and the general public about minorities and injustice. In practice, the UK corporate and national media appear to be much closer to Herman and Chomsky's (2002) view than Sen's (2009) ideal. This is evident in an London School of Economics study of the treatment by the print media of the former Labour leader, which is framed as a role of attack dog over watch dog (Cammaerts et al. 2016). Supporting these arguments was the Media Reform Coalition's report of the print media and TV news disinformation regarding anti-Semitism and Labour during the Corbyn leadership (Schlosberg & Laker 2018). It is also evident over the long term, with McArthur and Reeves (2019) conducting a quantitative analysis of UK print media finding a long history of this media demonising the impoverished, especially during economic downturns. Accordingly, Dorling (2020) argues that the mainstream media is more receptive to right-wing positions/parties. These parties/positions tend to have high social dominance orientation, or strong support for group-based hierarchies and injustices, while also correlating with climate crisis scepticism (Pearson et al. 2021; Milfont et al. 2018; Clayton et al. 2015). Thus, it seems fair to question Hayes and O'Neill's (2021) conclusions, or at least argue we need to combine an analysis of discourse with visuals to make any conclusions that the mainstream media is moving towards positions of intergenerational justice and equity.

Youth as Proxies

Also employing discourse analysis techniques, Knappe and Renn (2022) conduct an analysis of two youth groups active in CEC discourse: FfF and the UN's Major Group for Children and Youth (MGCY). They argue that these youth groups are in an open process of translation and politicisation – a "reinterpretation and appropriation" – of environmental justice, climate justice, and intergenerational justice. This reinterpretation and appropriation are motivated by the deep sense of injustice that young people will be much more affected by the CEC but have little agency in dealing with it (Knappe & Renn 2022, p. 4). The two groups produce varying discourse, with FfF being more pessimistic, informal, diverse, and presenting the CEC as something here and now and demanding solidarity with workers and communities most affected. The MGCY group presents a more formal, gradualist, and optimistic tone, with this difference, according to the authors, likely down to the cohorts. FfF is an open internationally diverse and fluid institution, while the MGCY is a more select, NGO-chosen few. However, both position themselves as proxies for future generations, with the MGCY explicit on this point, arguing "young people might only constitute 50% of the population now, we are 100% of the future," while FfF seems to perform this proxy role more implicitly (MGCY, 19/08/2017, quoted in Knappe & Renn 2022, p. 7). However, FfF is more explicit about the challenge of the power dynamics, as evident in their song sung to the revolutionary tune Bella Ciao:

And you continue
to feed the richest
the one per cent, one per cent, one per cent cent cent
You build the runways and burn the fossils
and let the planet pay.
But you don't realize
the cake's not growing
you take away, take away, take away way way.
You build the runways and burn the fossils
and let the planet pay.
Burn all the fossils
heads full of gossip
The lobby pays, lobby pays, lobby pays pays pays
You build the runways and burn the fossils
and let the planet pay.
You sit at tables
discuss the future
with greedy hands, greedy hands, greedy hands hands hands
We are the students who fight the system
because we have a world to win.
(Quoted in Spannring 2021, pp. 114–115)

The UN Convention on the Rights of the Child

Theoretically, Dryzek (2014; Dryzek & Pickering 2019) has generally called for institutions fit for the CEC that reflexively learn, adapt, and are democratically controlled and accountable. Lawrence (2021) makes the case that institutions for future generations should make use of proxies – current young people. van Dijk (2021a) explores cases filed for children via the Communication

Procedure and UNCRC and finds this institution being neither democratic nor adaptable and hamstrung by various other liberal nostrums. However, the cases are still of interest, as is van Dijk's work more generally. van Dijk treats future generations as youth now and unborn generations, thus tying in intra- and intergenerational justice. She also critiques human rights frameworks as perpetuating the "false dichotomy between human and non-human nature" and granting "massive power to states" (van Dijk 2021a, p. 2). This renders the human rights framework a poor tool for socio-environmental governance, pushing environmental legal scholars to focus instead on transforming international environmental law. The cases focused on exemplify these problems with 16 children from various majority world countries arguing actions by Argentina, Brazil, Germany, France, and Turkey violated these children's rights to life, health, and culture (van Dijk 2021a). These states refused to accept responsibility on a number of questionable grounds, the details of which are beyond the scope of this book.

However, van Dijk's (2021a, 2021b) broader conclusions and work are relevant. A significant conclusion is that the state is free to not sign and/or amend any agreement to its own liking. The Communication Procedure involves a general lack of transparency and high admissibility requirements, placing onerous time and effort burdens on the plaintiff. One particularly problematic condition is that a plaintiff must have exhausted avenues of redress in their own country before taking this case to the UN. This is time/resource consuming and problematic in repressive states that target those seeking justice. Another associated problem is that it is only states that can be filed against in this forum, not the problematic corporations, that in many ways are driving the CEC (Malm 2021; Koram 2022; Aronoff 2021). Despite the problems with this process, there were some direct and indirect benefits. Firstly, one argument the states tried to make was that plaintiffs not from their territory were not their responsibility. This argument was explicitly rejected, which could be seen as a step in overcoming the current political economy allowing the minority world to load emissions and toxic impacts onto the majority world (Táíwò 2022; Gelderloos 2022). Another less clear impact of this process was evident in a later case filed by Torres Strait Islanders against Australia. In this case, the Torres Strait Islanders called for the reduction of emissions as well as demanding $20 million in adaptation funds. Subsequently, the Australian government committed to $25 million in adaptation funds for these low-lying islands (van Dijk 2021a). There are open questions of whether this case led to this commitment and about what distribution this funding will take and how it will impact people. But it is a positive sign, nevertheless.

Thus, overcoming the supposed trade-offs between intra- and intergenerational justice involves listening to and learning from young people and their concerns. This is because these young people are a link to the future, and many young people understand this. The young today are victims of intra- and intergenerational injustice. However, they are also agents "who fight the system" and the best proxies we have for unborn generations and perhaps the best hope for transhistorical justice (Knappe & Renn 2022). In order to achieve transhistorical and trans-geographical justice, we need theoretical framework methods and tools to apply the framework. The next section will outline an updated form of the capability approach that offers a theory of substantive justice. The final section will examine a decolonised backcasting method that might help move us towards this substantive world of justice for all.

The Intergenerational Capability Approach

The capability approach was developed by Amartya Sen and Martha Nussbaum and has gone through many versions and adaptations. As shown in Chapter one, Page (2007c) chose this approach as his *what* of justice, while more recently Kimmerer (2013) and Winter (2022) also used this as part of it for their vision of indigenous intra- and intergenerational justice. The capability approach developed by van Dijk (2021b) will be the basis of this section. Crucially, van Dijk argues that future generations now and in the distant time are not and will not be a homogeneous group, with for instance able and disabled, gender, and indigenous groups amongst other differing groups. In this, she is purposely following Sen's (2009) normative point that justice should follow an equity principle that accepts people have different starting points and thus require differing treatment for similar outcomes. Robeyns (2017) labels how we change a means (food) to an end (nourishment) as conversion factors, and these can be personal (illiteracy), social (racism), or environmental (cold climate) and are significant factors on what people can do and be. van Dijk (2021b) goes on to outline a capability framework generated from her research on litigants to the Communication Procedure of the UNCRC (van Dijk 2021a), and empirical studies of human physical needs. These capabilities can be seen to correspond with Oliveira's (2022) sufficientarian approach and include *life and health, shelter and environment, bodily integrity, identity, expression and belief, knowledge and education, political empowerment, and relations to non-human nature and transcendence.* Table 4.1 gives more detail on this framework with some illustrative examples.

This list of capabilities developed by van Dijk (2021b) seems like a good place to start. Many in the world today lack these capabilities, so they can be seen as an intragenerational programme of justice – *the what* – to strive for, and one that synergises with intergenerational justice. It could also be achieved within planetary boundaries and so would correspond with sufficientarian approaches (Oliveira 2022) if economic growth for growth's sake is abandoned – as van Dijk (2021a, 2021b) suggests. There is one implicit trade-off within this capabilities framework and that is the issue of redistribution. For all to live with basic needs met and higher functions like political and social autonomy allowed to flourish, we need to redistribute

Table 4.1 Capabilities for Inter- and Intragenerational Justice

Capability	*Explanation*	*Example of Inter-/intragenerational Justice Issues Around This*
Life and health	Being able to live/enjoy a life of average length in good mental/physical health	Extreme weather events, food security, and infectious disease all set to exacerbate physical health issues. Also, the porous nature of physical and mental health is compounded by issues such as eco-anxiety and other environmentally associated mental health issues.
Shelter and environment	Being able to be housed and live in a safe and pleasant place	This is more than instrumental to life and health and is part of a person's sense of identity and empowerment. For certain indigenous groups (Winter 2022; Estes 2019), spiritual sense of place and connection to land are part of what defines them and creates links between the past and the future
Bodily integrity	Being able to be protected from violence, state, or otherwise	The CEC is set to exacerbate national security issues and interstate violence. Also, this crisis is set to exacerbate an already tragic cycle of violence associated with migration
Identity, expression, and belief	Being able to establish/express one's beliefs and live or not according to a religion	This may seem an issue future generations would be free from interference with; however, in certain situations, this is not true. Animist beliefs are dependent on species and land formations that can and are being made extinct or damaged/destroyed
Knowledge and education	Being able to be educated, to use and produce knowledge	This should include access to intellectual property and scientific research. Another issue is cultural imperialism, including chief amongst this form of oppression, the denial of genocide. This could pass into the collective memory of future generations undermining the credibility of the victim group and reinforcing one group's (the ethnoclass elite) domination through time
Political empowerment	Being able to take part and have a reasonable influence in political decision-making	Across the world women and other minority groups are poorly represented in policy-making. Women and certain indigenous groups are associated with more environmentally benign lifestyles and choices, and people involved in environmentally focused deliberation can become more environmentally aware and concerned. But beyond these instrumental reasons, political empowerment of all people is simply the just thing to do/have/maintain in a democratic society

(*Continued*)

Table 4.1 (Continued)

Capability	*Explanation*	*Example of Inter-/intragenerational Justice Issues Around This*
Relations to non-human nature and transcendence	Being able to live in/with and care for nature, animals, and plants, involving a more-than-human derived value/meaning	Indigenous groups have long advocated and protected the intrinsic value of non-human persons, but non-Indigenous groups can also have strong relations and interactions with their natural surroundings, and this has been shown to correlate with better physical and mental health indicators
Activities	Being able to engage in meaningful activities, including work and play in daily life	As the late David Graeber (2018) described, many people's jobs have lost meaning if they ever had any, and this is socially damaging, as is the inverted pay structure where the more socially useless and/or damaging, the higher the pay and vice versa. Davies et al. (2022) show how the pandemic had a photosynthetic effect – both highlighting and exacerbating injustice with record profits for the very richest in society and death and impoverishment for those on lower incomes and in key industries. Rushkoff (2022) describes an aspect of this as a dumbwaiter effect, whereby the more privileged were able to shield at home, while key workers who enabled this shielding and paid the tragic price were hidden from view by the tech platforms we use Looking into the future, certain jobs will not exist or be fundamentally changed by the CEC, and so justice demands changes to how time, labour, and resources are used
Social relations	Being able to form, develop, and maintain social relations	Living in the CEC now and in the future is placing and will place more burdens on those with less social support. Social cohesion, social trust, and building strong social institutions are the work of generations and so is the work of intragenerational justice leading to intergenerational justice.
Self-determination	Being able to live with autonomy including over time, geography, bodily, and socially	If an environment is unsafe, one should have the freedom to move, and this is one of Graeber and Wengrow's (2021) fundamental freedoms many societies of the past enjoyed, which many, especially racialised migrants, are denied today. Compare the terms ex-pat and migrant and the ethnoclass connotations of these (Goodfellow 2019)

(Continued)

Table 4.1 (Continued)

Capability	*Explanation*	*Example of Inter-/intragenerational Justice Issues Around This*
		In a just society, automation could be a force for good, with, as Keynes (1930) predicted, radically reduced working days/weeks. However, what Keynes did not really understand was the why of innovation. As Malm (2016a) shows at length during the United Kingdom's industrial revolution, innovation was about increasing social control and power of the individual capitalist and his class more generally. This, as Rushkoff (2022) argues, has morphed into The Mindset, whereby the tech billionaires create various damaging Ponzi schemes with the aim to get rich enough to escape the socio-environmental catastrophe their actions help drive Alternatives that respect autonomy abound with reduced working weeks, universal basic income and/or services (UBI/UBS), and democratic workplaces like coops, coop, and social/communal housing. Crucially, we need to challenge the logic of extraction that underpins our social system and is undermining most people's ability to socially reproduce now, and everyone's ultimately if the climate becomes too hot/unstable
Respect	Being able to be treated with respect and dignity, with no unjust discrimination	Again, migrants are treated terribly in the United States, the United Kingdom, and Europe, despite these countries/groups of countries being responsible for colonial extraction, neocolonial extraction, and colonisation of the biosphere, with these states responsible for an outsized share of cumulative emissions (Gelderloos 2022; Hickel 2020) Crucially, this capability would demand taking seriously the lives and livelihoods of young people today and future generations

Source: Adapted from van Dijk (2021b)

finite resources. At the same time, we need to reframe and refashion those resources that are made artificially scarce, such as political power or care. This all demands, as the IPCC recognises (IPCC 2022), a radical transformation of our societies. This leads to *the how*, a much more troubling and difficult question especially under the conditions of capitalist realism (Fisher 2009). Under this "realism," it is easier to imagine the end of the world than the end of this system, as depicted in much apocalyptic and dystopian science fiction. Perhaps one way of moving towards a more just future is to liberate how we imagine and project into this future. Currently, much of this work is done by groups working in the liberal and economic mode of thinking this book is arguing against. This results in the current scarce justice world being projected into the future. Thus, the final section will discuss how we might liberate our collective imagination in order to envision and build towards a more just world.

Future Studies, Decoloniality, and Backcasting

Future studies and scenario planning have long been used by businesses, most infamously by the fossil fuel industry in the 1970s. These companies predicted the CEC with remarkable accuracy, only to bury this research and spend large amounts of time and resources denying the CEC (Aronoff 2021). This culpability of the fossil fuel industry is part of a broader malaise of future studies that links to the arguments presented earlier when considering the limits and exclusions of liberal philosophy and mainstream economics. These are the core assumptions and practices of mainstream future studies that use concepts such as the four Ps (the possible, preferable, plausible, and probable), narrow epistemologies and circles of expertise, binary thinking, and other synoptic rationality tools that tend to reinforce existing power structures (Krishnan 2022). Unsettling this coloniality of foresight requires rethinking future thinking, but not in technocratic ways, but in democratic and justice-focused ways (Engle et al. 2022). This section will briefly outline some of the problems and oversights of liberal modes of thinking about the future, and the alternatives and correctives to these oversights and problems. It will then briefly consider backcasting in this light.

Mainstream Future Studies

Several critical voices call upon us to become good ancestors (Engle et al. 2022; Krishnan 2022; Táíwò 2022). Perhaps one way to answer this call is to specify in what ways current practices of future studies fail to do this. An initial problem is treating the diversity of human experience as a homogeneous, acultural "public," as much uncritical Anthropocene literature does

(Steffen et al. 2011; Lewis & Maslin 2015), technoscience forecasting does (Deloitte 2021), and much of the intergenerational justice literature dealt with in Chapter one does. This is strange as all of the recorded European history is driven by gender, class, ethnicity, and ability divisions and power relations, which is why these injustices exist now across the world it has created in its own image. Power dynamics will exist in the future; even in a near utopia, the question is how these are managed and the social relations these imply. One answer to this is the explicit acceptance of positionality of the people conducting future studies and consideration of how futures and trends play out in different groups and countervailing mechanisms, such as wealth taxes, to manage injustices (Krishnan 2022; Winter 2022).

A related issue is who is excluded from this future imagining. Presently, even in specifically oriented intergenerational institutions, there is little evidence of engagement with those set to be most impacted in the immediate future by the CEC (Jones et al. 2018). This includes not only children and young people but also those groups routinely misrecognised or not recognised by our ethnoclass, gendered, and ableist polities. Deliberative and bottom-up practices and institutions might be slow and difficult. However, these practices are able to include and listen to a wider variety of voices, including the signals of our natural world, and so are worth the time and effort (Dryzek 2013). Also, if we are thinking of deep time, then time is not something we lack if we can collectively mitigate the CEC or the worst of this. In contrast, functioning and deliberative forums that have the power to make things change are rare. For instance, the CoRWM process, while laudable and one of the most democratic exercises in UK history, which is admittedly a low bar, was still characterised by experts making many of the key decisions (Chilvers & Burgess 2008). Thus, a move from participatory engagement to one of participatory decision-making is due (Krishnan 2022). This is because, fundamentally, behind/above all technoscience questions of *what* and *how* are the normative questions of *why* and *should we do this*, and it is scientistic and undemocratic to deny this (Fischer 2000; Habermas 1996). Finally, the idea of ethical expertise is a strange one as the movement from morality to ethics is one of consideration and deliberation. This implies the aim should be to build collective ethical consciousness through deliberation.

A further issue surrounds the concept of temporality. Time in the liberal tradition is linear, with, as all intergenerational justice theorists recognise, the present elevated as the most integral and important time. This has varying impacts with, at the sharp end, many economists advocating and governments implementing social discounting as if this was a simple prudent choice, and not as Caney (2014) argues akin to racism or sexism. More generally, this notion of linear time connects to the idea of progress or what Appadurai (2013) names trajectorialism, or Western modernity's narrative trap. In this linear, cumulative journey from here to there, now to then, humanity and all its kaleidoscopic differences are collapsed into homogeneity, a monolithic people that moves from cave to farm, from monarchy to democracy, with the ultimate teleological destination looking like Europe or the United States (Graeber & Wengrow 2021; Appadurai 2013). With the socio-environmental damage mounting as this myth of progress fades, Walter Benjamin comes to mind, with his idea of the angel of history. In interpreting a Klee painting, "Angelus Novus," Benjamin (1968, pp. 257–258) sees the angel of history not observing a chain of events but "one single catastrophe . . . piling wreckage upon wreckage" at his feet. The angel wants to "make whole what has been smashed" but is caught in the storm propelling him into the future – "[T]his storm is what we call progress." Or more prosaically, as Terry Pratchett (1991) wrote "Progress just means bad things happen faster."

The way out of this storm, for Benjamin (Benjamin quoted in Sperber 2018), and the degrowth movement (Hickel 2020) is perhaps the emergency brake: "Marx said that revolutions are the locomotive of world history. But perhaps things are very different. It may be that revolutions are the act by which the human race travelling in the train applies the emergency brake." What, one may ask, is this emergency brake? Partly, this could be seeing time as, in an important sense, cyclical like the seasons, night and day, or human and animal lifecycles, or even capitalist accumulation/appropriation cycles. Arguably, doing so holds a number of advantages.

In a broad sense, unhooking time from the idea of linear progress and eternal motion might allow the more sustainable notions of steady state economics to gain traction over those of endless dynamic growth (Daggett 2019). More specific to intergenerational justice, time as circular/spiral, as animist traditions and indigenous theorists frame it, allows a more direct present relation to the past and future (Watane 2022; Winter 2022; Estes 2019). Seeing the present time as part of an arc or loop leads to conscious and careful reference to the past and similar consideration of future generations as heirs of one's actions in the present (Watane 2022). Thus, this understanding of time can lead to reflexivity where "past and future are intimate bedfellows with the present" (Winter 2022, p. 18). This spiral or circular thinking also escapes the specious and convenient forgetting (Patel & Moore 2020) and strategic ignorance (McGoey 2019). With technologies such as nuclear, at first this involves deep uncertainty surrounding initial design and efficacy and involves mistakes, some of which are overcome substantively and others rhetorically. However, all this becomes obscured and sanitised by a linear narrative of technological progress (Verma 2021). More broadly, this ignorance and forgetting are a result of teleological approaches to history, whereby the end justifies the means, regardless of what was done to whom and how. Appended to this is the idea that we are starting from endless baselines, not reinforcing and repeating past injustices.

A final but perhaps overarching issue of future studies is the ontology underneath. This can be labelled liberal/rationalist and has been critiqued throughout this book. To recapitulate the broad problems of this ontology:

- It is hierarchical with these hierarchies built and based on property/asset ownership and involving marginalisation and oppression of specific groups;
- It is individualistic with the main social unit the individual and the focus on this individual's negative freedom *from* external constraint (even though many marginalised people are extremely constrained in practice), rather than the positive freedom *to* do or be (to eat, for example);
- It is anthropocentric treating the world instrumentally, as input and output, and it is based on binary thinking and a reductive science that is a poor model for our connected world.

As a counter to this liberal rationality, Winter (2022) proposes a cosmological worldview entailing humans as part of Kimmerer's (2013) democracy of species, including non-living things like rivers or mountains. Justice for humans depends on justice for all the living and non-living, as we are part of a web of life, and nature is everywhere and in everything, and so "dignity and integrity inhere to everything and to the whole" (Winter 2022, p. 6; Moore 2015). With these considerations, what is backcasting and what would it look like done in this justice-centric and decolonial way?

Backcasting Decolonised?

Backcasting is a way of thinking about the future but differs from forecasting methods in future studies that aim to maximise profits by aligning corporate action to likely events (Robinson 1982). Backcasting by contrast fronts the normative element of future thinking, asking what future we desire in terms of justice and fairness, and what would need to change to achieve this future (Cotton & Rowe 2019; Cotton 2013). The idea of backcasting emerged partly from Robinson's (1982) research on energy and concern with the technocratic forecasting of future studies that ignored the normative "soft energy" pathways of renewable energy. Backcasting splits futures into "probable" – most likely in participants' view; "possible" – less likely; and "preferable" – the most desirable and ethical (Cotton & Rowe 2019). A decolonial perspective would add the P of "perspective" or whose perspective is being expressed and/or privileged (Krishnan 2022). For instance, most deliberative exercises exclude children and younger people, so an adult, and thus arguably more presenteeist, perspective is being expressed and privileged in these exercises.

Methodologically, backcasting attempts to bridge the psychological distance involved in consideration of futures remote in time, space, and imagination (Cotton et al. 2019). To achieve this, it is helpful to look backwards first before looking forwards to build context (Cotton 2013). The question here is how far and wide one does this, but this is also a question going forward, or regarding how far the exercise projects into the future and what issues to cover (Robinson 1982). Thus, a backcasting method design might have three features: a review/discussion of past events/trends; an exploration of probable, possible, and preferable futures; and some limitations of spatial, temporal, and ideational scope (Cotton & Rowe 2019; Cotton 2013). This could involve various deliberative methods; however, of interest to this book is the Deweyan method as described broadly by Cotton (2013), in which are the twin aspects of building moral empathy by placing oneself in another's position and a dramatic rehearsal of different futures and how to approach/achieve desirable ones. An example of this might be writing to or from the future, as they did in the intergenerational dilemma studies. The objective of these exercises is to "imagine complex, 'thick' descriptions of possible futures, involving competing possible lines of action and courses of conduct, followed by a deliberative process involving imagination, creativity and emotion" (Cotton 2013, p. 323). These approaches recognise that moral empathy is like a muscle needing exercise, and this requires the imagination to extend one's understanding beyond one's narrow field of experience and that this is best done socially and deliberatively. This is what Galvin (2020; Schlosberg 2013) is suggesting when he argues that justice is not some eternal given but something we build and negotiate in our social interactions.

Finally, a decolonial aspect of this method would involve design, method, and outcome features. In design, it would follow the most recent innovations in deliberative forums, involving

- sortition in a non-random selection so as to achieve a representative sample, crucially including those usually excluded, such as children;
- good deliberative conditions providing time, expenses, and respect, including space for emotions, storytelling, and other less traditional forms of "rational" deliberation (Young 1996), while avoiding other issues such as group thinking or false consensus (Cotton & Rowe 2019); and
- final recommendations by super majority of >70% or consensus in order to build common ground and move beyond zero-sum, trade-off thinking (Cesnulaityte et al. 2022).

Methodological focus would be on extending the geographical and temporal scope. This expanded scope would explore how our energy and social systems are mired in historical and ongoing colonial/neocolonial and anti-democratic oppression and violence (Estes 2019; Powell 2018; Voyles 2015; Mitchell 2013; Moore 2015). In terms of outcome, there are two issues. Firstly, these should not be stand-alone events but, as far as possible, ongoing with follow-up events and evaluations, not just amongst the experts but including the participants (Cesnulaityte et al. 2022; Davies 2007). The aim of this continuous process is to develop the democratic, moral, empathic, and imaginative capabilities of the participants. If this ongoing deliberative process was expanded to more areas and issues, it could help redesign and revitalise our historically limited and contemporarily attenuating democracies (Dryzek 2000; Williams 1989; Quilter-Pinner et al. 2021). Secondly, and perhaps even more challenging, is to make these forums count as decision-making bodies, for deliberative exercises that are undermined, misrecognised, and dismissed is corrosive to democracy. This can be seen in the growing distrust in politicians and in our largely performative democracies (Quilter-Pinner et al. 2021: Graeber & Wengrow 2021).

Conclusion

The overall argument of this chapter was that today's children and young people will face far more challenging aspects of the CEC than those currently in power across society. These policy-makers and adults today have at best abdicated responsibility (with some notable exceptions), at worst been captured by big finance and the fossil fuel industries. Thus, a just and democratic aim should be to include these children and youth not only as proxies for future generations but also as political actors in themselves. Additionally, we outlined an adapted form of the capability approach that would serve as a substantive and inclusive model of justice to aim for tying intra- and intergenerational justice together. We then suggested a method of backcasting that could help us move towards a more just future. Together these new actors, new thinking (although much of it leaning on older ideas), and new deliberative praxis could take the first steps on this journey.

This journey begins with understanding that democracy was never given or linearly progressed to in some form of social evolution. Rather it was struggled and fought for at different times and in different places, often far from parliamentary institutions. Energy and its materiality has long been part of this struggle (Cotton 2017; Malm 2016a; Mitchell 2013; Merchant 1989). This struggle has never ended for certain social groups within the minority world and across the majority world. This struggle will continue and is the subject of part two of this book, which examines nuclear waste cases.

Part Two

Nuclear Waste and Intergenerational Democracy

5 Critical Nuclear Concepts

This chapter examines the critical theories that have been used to describe how the nuclear affects the socio-environmental world. These effects are significant and complex involving a globe-spanning spatial imprint that is seen, amongst other things, as a "golden spike" signifying the Anthropocene, and temporal aspects that enter the geological scale time frames. Thus, the nuclear is a core aspect of intra- and intergenerational justice. More specifically, this chapter examines the notions of nuclear landscapes and communities, peripheralisation, energopower, and nuclear colonialism. These concepts are crucial in understanding the injustice/justice of the nuclear and will be helpful in the chapters that follow examining specific country case studies of nuclear power and waste.

Nuclear Landscapes and Communities

Firstly, the idea of nuclear landscapes describes how the nuclear pervades socio-political landscapes and reshapes these in various ways. At its most dramatic and destructive, this includes not only the nuclear attacks on Japan but also the numerous nuclear bomb tests conducted across the globe (Alexis-Martin & Davies 2017). More prosaically, it defines a "confluence of physical and social manifestations of nuclear objects, activities, and relations embedded in context, producing a multiplicity of stakeholder relationships, multiple senses of place, and unique power relations" (Bell 2022, p. 3). Landscapes in this sense are multi-scale, connecting the geopolitical from the international to ultimately the local; multi-relational, involving socio-economic, political, and more direct relationalities to land; and spatio-temporal, allowing for understanding how energy infrastructure, especially the nuclear, has its spatial imprints but even larger temporal writing upon landscapes and their history (Bell 2021, 2022). The idea of the nuclear community emerges from this nuclear landscape and involves the economic power of the nuclear energy infrastructure in a particular community where it often becomes the major employer and is highly involved/integrated directly and indirectly in the local

DOI: 10.4324/9781003422686-8

economy (Bell 2022). This idea also encompasses the technoscience and engineering expertise that develops in tandem with the nuclear industry and forms a supportive and authoritative community of advocacy (Szulecki 2018). This economic power, gained via privatisation of public assets usually for a fraction of their market value (Christophers 2020), together with the associated jobs and expertise, functions to develop "relations of familiarity, knowledge, and trust with members of the public" (Bell 2022, p. 8). However, as with anything social, this is not so settled and one-way, with distrust and opposition emerging partially due to the injustices associated with the troubled history of nuclear weapons and power (Flynn 2003).

Peripheralisation

Peripheralisation is a concept that emerges from Wallerstein's (2006) and others' work on world systems theory, in which the minority world imperial core of Europe and the United States, via asymmetrical political, economic, and military means, extracts value from the majority world. However, this concept has been fruitfully applied by Blowers (2010) to processes within states in the minority world, specifically the United Kingdom. In this national–regional–local sense, there is the urban–rural and North–South divides, whereby the power-intensive urban core takes the benefit of electricity without facing the costs of the locally, peri-rural emplacement of nuclear infrastructure. Blowers and Leroy (1994; Blowers 2010, 2020) identify five aspects of peripheralisation:

- *remoteness* or geographically distance or inaccessibility from the urban centres;
- *economic marginality* characterised by monopsony power or one or few employers/taxpayers in the area able to monopolistically dictate terms and conditions (Christophers 2020);
- *political powerlessness*, where crucial decisions affecting the locale are made in distant urban and economic cores;
- *cultural defensiveness*, where locals express ambivalent fatalistic acceptance of nuclear infrastructures; and
- *environmental degradation*, where areas either are radioactively contaminated or face higher risk profiles.

This idea of peripheralisation crucially fleshes out the notion of nuclear landscape. This framework enables the exploration of how the above conditions or aspects of peripheralisation interact, reproduce, and reinforce each other in these nuclear communities. It also enables the exploration of how communities resist and organise against these spatio-temporal injustices (Blowers 2010, 2020).

Energopower

Energopower is a concept that sees how electric or fossil fuel energy/power reproduces and reinforces power in the second social sense. Foundational thinkers in this space were Nader (2004) and later Mitchell (2013). Nader (2004) worked with the nuclear natives and energy specialists of the National Academy of Sciences CONAES Project in 1970, but as an inside outsider, being the only anthropologist and woman amongst around 300 specialists. Here, Nader (2004, p. 773) found "unscientific attitudes in scientific places" and a more general path dependency compounded by a number of aspects. Nader (2004) details this involved the TINA (there is no alternative) arguments about nuclear power, which have been, at least in part, resuscitated by climate change security discourses (Cotton 2020; Blowers 2010). There was also capitalism's core assumption of perpetual economic growth linked to the assumption of ever-rising per-capita energy consumption (an aspect of unscientific science). This growth assumption was in turn connected to the conflation of technological progress with social progress. Finally, these energy experts used Haraway's (1988) god trick by imagining themselves as disinterested and objective actors standing outside the system, not situated humans and part of it. Indeed, Nader's commentaries that described science bureaucracies as exerting power and explored these bureaucracies' limitations in how they define and propose solutions, but also in how they frame the broader cultural outlook, were adamantly opposed by these bureaucracies. In the god trick mode, these bureaucracies maintained "science is autonomous and culture free" (Nader 2004, p. 776).

Mitchell (2013) examines the materiality of fossil fuels and how this shapes, allows, and limits democratic functioning. He details how coal miners, rail haulers, and dockers were able to halt the movement of coal and stall the national capitalist economies. In doing this, these workers extracted the democratic reforms that accumulated into modern representative democracies. This is a story of social struggle, with political and economic elites almost universally opposed to democratic reform but forced by events into compromise, control, and limitation of these democratic experiments (Graeber 2007). One aspect of controlling and limiting this democracy was oil and its materiality and deployment. Crucially, this transition[1] to oil was not cheaper during the key period and was more about shaping and engineering society to limit democratic power (Bonneuil & Fressoz 2016; Mitchell 2013; Huber 2013). This can be seen in two parallel but complementary processes – the suburbanisation of US urban spaces and the Marshall Plan.

Huber (2013) documents how the US governance and oil interests collaborate to form, reinforce, and reproduce a sociotechnical imaginary and materiality in the United States. This is predicated on the lifeblood, oil, and connected to an entrepreneurial life, where one identifies one's life with car, suburban house, and the energy-intensive things within, thus supporting individualism and consumer capitalism. However, the brave new suburban world was not open to all, with ethnoclass exclusions being part of the overall design (Huber 2013). The United States' Marshall plan aimed to rebuild Europe after the Second World War, but any rebuilding exercise will have normative aims, and this was no exception. Mitchell (2013) details how this had three broad features. Firstly, it aimed to impose in Europe North American ideas of labour relations and industrial management, which involved more supervision and working more for less. The second condition was economic integration that was exemplified by the early European Union institution, the European Steel and Coal Community. The emergence of this institution and this economic integration undermined unions through reduction of their numbers and collective power via cross-border trade of coal. Thirdly, and most significantly, this programme aimed to move Europe's energy system from one reliant on coal to one reliant on oil. This involved funding refineries, road infrastructure, and car manufacturers all using oil secured from the Middle East, controlled by US oil firms who only accepted US dollars. This resulted in the emergence of the petrodollar system through which the world pays tribute to the United States for the energy system it chose (Mitchell 2013).

These ideas of energopower can and have been applied to other forms of power. Cotton (2017, 2020) does not use the term energopower but does argue that the UK government, when it realised it could produce electricity as a by-product of its nuclear weapons programme, used this form of power politically against the mining unions. More theoretically, Burke and Stephens (2018), following Winner (1980), look at technology, and specifically energy technologies, as embedding political values to differing degrees. This involves how concentrated energy sources and infrastructure tend to concentrate political power as in our world. In contrast, decentralised renewable technologies such as wind and solar, while not ensuring more democratic formations, at least allow for these formations (Burke & Stephens 2018). Thus, as Mitchell (2013, p. 8) argues, political possibilities are "opened up or narrowed down by the different ways of organising the flow and concentration of energy," which are "enhanced or limited by arrangements of people, finance, expertise and violence" in energy assemblages. To exemplify the energopower of the nuclear, the next section will summarise aspects of what has been described as nuclear colonialism.

Nuclear Colonialism

The United States has a long and terrible history of genocidal treatment of the Native Peoples (Edwards & Kelton 2020). This brief section will explore the history of the Navajo Nation, uranium mining on Navajo land, and uranium mining more broadly. Curly (2014), reviewing the history of this nation from 1868 to 1937, makes clear that, in Scott's (1999) seeing like a state sense, the United States made the Navajo Nation in order to make the Navajo people "legible" and thus subject to simplification and exploitation. For instance, this operates through imposing male-centralised leadership, or an "indirect rule" model taken from British colonial officials in Africa – and then manipulation of this leadership by both the US state and powerful interests within this. Curly (2014) emphasises that this "indirect rule" model was imposed on the Navajo population, who had been subject to the infamous Long Walk from their lands to internment in New Mexico, where a much-reduced, half-starved, and desperate population managed to get permission to return to their lands. However, these lands' borders and the conditions of this return were dictated by the colonial state. Overall, this is an egregious cosmopolitan injustice as well as an injustice of process, distribution, and recognition, and part of a much wider genocidal period (Dunbar-Ortiz 2023; Edwards & Kelton 2020). Powell (2018) agrees that this political entity was largely a creation and creature of the coloniser but stresses that this entity was created with the aim of energy extraction, a result of a broader accident of colonial planning, namely placing "Indian Territory" over a large proportion of the geological mineral resources. Both Powell (2018) and Curly (2014) emphasise, however, that the Navajo people resisted these processes of colonial simplification and manipulation, for instance through settling beyond imposed borders, openly defying both federal and tribal authority, and resistance to energy extraction. This mode of seeing like a state (Scott 1999) can be seen as fundamentally an injustice of misrecognition, for it creates units and subjectivities that did not exist previously, as in the tribal leadership, or removes or occludes these subjects, as in the trans "two spirit" people who were a crucial part of pre-colonial indigenous societies (Estes 2019; Powell 2018).

Powell (2018) describes the history of this energy extraction moving through oil, uranium to coal, with much of this used to power the urban sprawl of the "Sun Belt." This constitutes clear injustices in the distribution of energy wealth and illth. Although there was some electrification within the Navajo Nation, this was not a priority with many places deemed too "rural." Crucially, before the uranium mining phase, the colonial state turned its eye upon the Navajo Nation's sheep. This was connected to the process of building the Hoover Dam, a project pushed by and named after the man who did much of the crucial work against the world decolonisation movement (Harris 2023). New Deal planners believed that these sheep and their grazing would lead to run off and premature siltation, thus reducing the dam's power output, a threat to modernisation. However, these sheep for the Navajo women, who largely managed them, were less a commodity but more of a companion species in Haraway's (2016) sense, or perhaps in the broader indigenous sense of a democracy of species and the web of life (Winter 2022; Kimmerer 2013). From 1934 the colonial state tried to tax the sheep, which failed largely because the sheep "ownership" was not individualised – a further example of misrecognition. The state then turned to extermination, a tactic it had used earlier on Lakota people and the plains buffalo (Estes 2019) and the Navajo people's sheep and dessert orchards and agriculture in the 1860s (Voyles 2015). In a twisted "ecological" reprise, the sheep herds were slaughtered in tens of thousands, devastating the Navajo families who relied on them materially and culturally. As a result, many families were forced into the wage economy, and the wider Navajo Nation was desperate for investment when it came calling (Voyles 2015; Powell 2018).

Uranium mining began on Navajo Nation land in 1942, with a rich seam of this resource within Navajo lands. Powell (2018) describes how 340 million pounds of U3O8 (uranium compound) have been extracted, half the United States's domestic reserves between 1942 and 1980s (Voyles 2015), but that 400 million pounds remain in the ground. However, the contracts for the uranium mine concessions were often well below the market value for the mineral (Voyles 2015; Endres 2009), likely because after the impoverishment caused by sheep extermination, the tribal leadership was desperate (Powell 2018). In addition, it appears the Bureau of Indian Affairs (BIA) negotiated these contracts in a one-sided way, ensuring that Indigenous Nations only received around 3.4% of the value of the mineral. This bad faith from the BIA speaks not only of injustices of the process but also of how this institution's role was seeing like a state – not only to make natives legible as waged workers but also to exploit through the lowest paid and most dangerous work (Gómez 2022; Voyles 2015). In contrast, firms doing uranium prospecting, mining, and milling, along with the rest of the nuclear industrial complex, enjoy generous tax breaks and subsidies, without which this complex would likely fail (Koplow 2011; Bridle & Attwood 2016). More broadly, Voyles (2015) describes a uranium legacy of around 2,000 abandoned mines sitting open or poorly covered or marked, with tailing piles littering the landscape leaching radon gas, as well as other toxic contaminants, into air, water, and the broader ecosystem.

This is a history that began with promise – like the broader tech optimism period of the early nuclear era – but then develops into all too predictable and tragic consequences. Powell (2018) and Voyles (2015) report that cancer rates amongst the Navajo were, prior to uranium mining, so low there was talk of immunity amongst medical researchers. However, by the 1980s, this picture had radically changed with cancer rates amongst miners and those living nearby skyrocketing and sharp increases in breast and ovarian cancer and birth defects, along with a host of other known and unknown serious and fatal conditions. In addition to being given the most dangerous jobs for the least pay, Navajo miners were not informed of the risks or given protective equipment, despite these risks being scientifically and medically accepted at this time (Hecht 2012; Voyles 2015). There was widespread contamination of water in a water-scarce area, affecting people and livestock alike (Voyles 2015). These risks were further exacerbated by a widespread practice of building residences with contaminated material.

In 1979, on the border of the Navajo Nation at the Church Rock uranium mine, a tailing disposal pond breached its dam. This resulted in more than 1,000 tons of radioactive waste and associated contaminants pouring into the Rio Puerco River, in the largest spill of radioactive waste in US history. These contaminants flowed downstream through the Navajo Nation, but very few people were warned, mostly because the warning was not delivered in the Navajo language. Subsequently, the Navajo Nation called for federal disaster relief; however, this was denied. Overall, Voyles (2015) frames this as a form of slow or delayed violence due to the lag in the effects of exposure. Attributing this slow violence is made very difficult by the boom–bust nature of the uranium mining industry, where companies who employed miners often no longer exist. Thus, the impacts and clean-up responsibility fall on the Navajo Nation, the EPA, and the federal government. However, the federal government only became involved after decades of campaigning and provided an underwhelming response (Voyles 2015). Ultimately, the impacts fall on the women as the socially constructed care provider dealing with wreckage left behind. It is a testament to the strength of these and other racialised women that they formed the backbone of the emerging environmental justice movement (Harris 2023; Voyles 2015).

Finally, the international history and processes of uranium mining provide an overall example of cosmopolitan injustice and nuclear colonialism. Generally, as Hecht (2012) argues, uranium mining industries, like the broader mining and fossil fuel industries themselves (Harris 2023; Mitchell 2013), practise racialised hierarchies and operate in racialised geographies. This involves elite ethnoclass management and engineers directing operations and receiving the lion's share of wage/stock remuneration and racialised, lower-class groups doing the more dangerous mining tasks for little return (Harris 2023; Mitchell 2013; Voyles 2015). It also involves the broader divide between the high-tech nuclear applications in the minority world and the low-tech and dangerous mining in the majority world. Again, this distinction includes or is complicated by the internal colonised peoples of settler colonial states (Harris 2023; Powell 2018; Hecht 2012). This brief section shows how uranium mining tends to breach all four tenets of environmental/energy justice and continues to have a distinct element of coloniality (Sovacool et al. 2023; Bell 2020; Voyles 2015).

Conclusion

This chapter has summarised critical theories and concepts that have been used to describe the nuclear. These effects span from the global to the local and the now to the deep time future. Thus, they are issues of intra- and intergenerational injustice/justice. We first outlined the idea of nuclear communities and landscapes, which describes the manifold social, environmental, and political effects the nuclear has on our world across space and time. Secondly, we outlined the theory of peripheralisation, which adds detail to the ideas of nuclear communities and landscapes, describing how this involves geographical, economic, and political marginalisation. It also involves environmental degradation and a cultural ambivalence and ambiguity. Next, we outlined the theory of energopower, which describes how all our politics are plugged in and powered by material infrastructure that can concentrate and reinforce, as well as distribute and share, social power. Finally, we examined the idea of nuclear colonialism, or how this form of energopower has been used to oppress, appropriate, and destroy the lives and land of colonised people. This nuclear colonialism also describes the minority and majority world divides, where nuclear power and applications are used in the former, while the dangerous mining of uranium occurs in the latter but with few of the benefits.

These concepts will be used to assess the three cases that follow. The CEC will involve difficult choices and trade-offs in our energy choices, as all have impacts and carry risk. However, the argument of this book is that the best and fairest way to manage these technopolitical questions is deliberatively and democratically. As we will see in the next three chapters, democracy has not always been the nuclear sector's strength. In a recent nuclear conference, one industry actor summarised this tension claiming the industry had "allowed too much democracy" to impede the industry and that this required "hacking some of this nonsense out of the front and actually making decisions rational" (Lehtonen 2022). While this might be a quite extreme position that was also challenged at this conference, it does signal a problem that has long been associated with the nuclear. The democratic credentials of the nuclear is also a theme that connects all three cases and chapters that follow.

Note

1 It is more accurate to call this an *addition* rather than *transition* as more absolute amounts of coal tended to be consumed in any year up to 2019.

6 Canada and the Nuclear Waste Management Organisation

This chapter will describe the history of nuclear weapons/power in Canada and then move on to how nuclear waste is being managed. This management involves overall aims and objectives and the implementation of these aims and objectives. However, it is important to see these processes as distinct as there is always contingency and slippage between governance aims and objectives and place-based implementation. This is because governance is never a straight-forward process or cycle of conceiving, planning, and implementing, but a far more iterative and recursive process (Cotton 2017). Also, as Bickerstaff (2012) argues, whenever a government, as in the United Kingdom or Canada, conceives a shiny new plan, there is an implicit exorcism of the sins and mistakes of the past and an attempt to present a new beginning, or baseline. This is an aspect of forgetting and presenteeism this book has encountered again and again: however, in peripheralised communities, history and remembering are fundamental to identity and acts of agency and power.

Context and Histories

Headline figures for Canada's nuclear complex give some indication of its place in the world of nuclear energy. Canada has 19 operating nuclear reactors producing 13.6 GWe, or approximately 15% of total electricity in 2022, with 18 of these in the Ontario region, making Canada the sixth highest user of nuclear power in the world (WNA 2024). However, it is in the mining of uranium where the Canada story is more interesting and troubling. It has approximately 564,900 tonnes of uranium reserves, making it the third largest resource holder at 9%, but was the second largest exporter of uranium until 2019 (WNA 2023a). This story of mining and export of uranium in Canada, a former colony of the United Kingdom and settler state, has many beginnings and middles but no ends. Yet the Manhattan Project and the nuclear bombing of Hiroshima and Nagasaki are one significant middle (Van Wyck 2010).

This story begins on appropriated land in the far north of Canada with Eldorado Gold Mines (referred to as Eldorado) as the "central corporate protagonist" in a "very Canadian exercise in forgetting" (Van Wyck 2010,

DOI: 10.4324/9781003422686-9

pp. 24–26). In 1930, a sliver and mineral mine was discovered that had one of the world's largest radium deposits, with this resource in its first incarnation seen as a highly valued commodity and substance of wondrous medicinal power (Van Wyck 2010). This changed during the Second World War, with Eldorado being "transformed and fortified as a corporation of the Crown" and radium, as closely associated with uranium in location, becoming a military and national security concern (Van Wyck 2010, p. 26; CNSC 2012). During the war, the United Kingdom desired a partner to move its Cambridge-based nuclear laboratory nearer the United States, allowing the United Kingdom to work on the nuclear weapons programme, to which the Canadian government agreed, establishing Montreal Laboratory (CNSC 2012). This laboratory would work closely with the Manhattan Project in establishing what Van Wyck (2010, p. 36) calls the "Highway of the Atom," which, like many aspects of the seeming modern infrastructure, is really a "palimpsestic economic and colonial path used by Europeans carrying furs, food, and disease, and subsequently carrying radium and uranium." This is a highway in multiple senses. It is an A to B on a map from Port Radium in the far north of Canada to Edmonton, United States; it is also the journey of capitalism's "atomic diplomacy," which was arguably more about showing the "West's" power and preventing the share of war spoils with its erstwhile Soviet ally than a military necessity (Yagami 2009). And it returns to Great Bear Lake communities as "cancers, stories, addictions, and depression; hauntings" (Van Wyck 2010, p. 37).

The colonial-inflected injustices of this period of nuclear history are many. Yet these are somewhat occluded as it was a period of discovery and early part of the techno-optimism and secrecy/security that characterised early nuclear history (Cotton 2017, 2020). However, as this was never *terra nullius*, or empty land, the treatment of the colonised Dine people deserves remembering, not least because colonised people are often, and in this case as we will see, peripheralised communities. It is unclear whether any Dine worked in the mine, but they did perform many support roles to this mine, and their surrounding land was contaminated. This included Eldorado dumping some 1.7 million tonnes of tailings into the Great Bear Lake, arguably resulting in high cancer rates within the community (Van Wyck 2010; Henningson 2006). A further injustice to this community was finding out what this material was used for in Nagasaki and Hiroshima. This resulted in a delegation from the Dine people going to Japan to apologise (Van Wyck 2010). This apology is something Truman, the president to order this bombing, signally refused to do (Mitchell 2020).

These injustices were suppressed for a long time. The company commissioned history of Eldorado – in many ways the official history for a long time – was derided by a reviewer as an extended piece of public relations, and completely omitted any Dine involvement (Van Wyck 2010; Howard 2019). However, even when Canadian authorities began to recognise that Dine communities may have been harmed, this was at best grudging and keen to deny as much as accept wrongs. The Canada–Déline Uranium Table (CDUT) was set up in 1999 with the aim to "resolve the human health and environmental issues surrounding the Port Radium mine" and was made up of a range of Canadian settler officials and experts, and also Déline community representatives (CDUT 2002, p. ii). A first meeting in March 2000 established the "common language," or the epistemological foundations, that the process would be based on. This included three aspects: Traditional Knowledge – or the need to proceed with studies that "integrate scientific and traditional knowledge"; Fundamentals of Radiation – or the basics of types and routes of radiation exposure and linked risk of developing cancer; and Environmental and Health Risk Assessment and Communication – formal risk assessment (FRA) based on current scientific knowledge in order to minimise socio-environmental harm and the methods of communicating this information (CDUT 2002, pp. 1–2). However, as Cotton (2017) argues, assessing nuclear risk is a fraught process that cannot be reduced to quantitative FRA and cost–benefit type assessments. This is an argument that applies more broadly to the assessment of risk where experts decide the acceptable risk level for some lay communities (Fischer 2000; Jasanoff 1999). Although Traditional Knowledge in acknowledged in this framework, there are, as we will see, fundamental tensions between scientific and Traditional Knowledge.

In 2005, the CDUT produced a final report and recommendations, which exemplify some of these ontological and epistemological tensions, as well as Nader's (2004) contradictions within scientific discourse. A first point to make is that the CDUT (2005), similarly to Van Wyck (2010), was refused access to much of the relevant archive concerning the Eldorado mine, with no explanation given. How a process of healing can commence when the administrative state withholds much of the information from the crucial 1930–1960 period is a question of epistemic justice that arguably undermines the whole process. More explicitly, a number of the CDUT's conclusions were problematic. Firstly, the high cancer rates amongst the Déline community were disputed by the CDUT, which argued these rates were not statistically significantly higher than in surrounding areas; however, this report does accept that this statistical significance means little as the Dine sample size is too small and the locale's records did not go back beyond 1990 (CDUT 2005). The report also points out that when physical health problems among the Dine were reported, this was often dismissed as old age, and speeds of referrals for diagnosis and detection of cancer were slow (CDUT 2005). This is arguably an example of Nader's (2004) unscientific science, in that accepting the conclusion of a test when its assumptions are violated is a basic error, which is then compounded by the poor medical treatment and data that were available. More broadly, this is arguably part of what Hecht (2012) describes as a postcolonial divide whereby colonised others are exposed to higher risk or nuclear colonialism (Gómez 2022; Voyles 2015). Uranium atoms decay into radon, which if inhaled settle in lung tissue and expose it to alpha particles, but this process is inherently stochastic, varying with conditions in/around the mine, length of exposure, and other factors. Radon exposure also takes 10–30 years to play out. Therefore, with colonial legacy health systems, as in the Dine case, and epidemiology and animal testing, even when available, inconclusive and easily questioned by corporate lawyers, there emerge places in Northern Canada or states in Africa where these higher risks are common but harder to perceive and act on (Hecht 2012). This report did accept that the Dine community suffered fear and anxiety due to the sense of contamination from Port Radium, and thus this damaged this people's sense of harmony with nature, a core part of the cultural identity (CDUT 2005). Overall, there is an incoherence here with even the scientific fundamentals looking suspect and much of the key data being withheld. At the same time, the Traditional Knowledge, presented as the first part of knowledge in the framework above, becomes an aspect of perception and oral history kept separate and secondary to the scientific data.

The reception of the report was interesting and takes up many of the tensions between the different knowledges mentioned earlier. In the documentary (Henningson 2006) *Somba ke: The Money Place*, two interviewees make interesting critiques. Dr Rosalie Bertell argued that the report's focus on deaths undermined the study and that blood and urine samples would have been "simpler, more direct and cheaper" than the engineering firm, Senes, the CDUT chose as consultants. The narrator of the documentary adds that Senes had a conflict of interest as a member of the nuclear lobbyist the Canadian Nuclear Association. The Journalist Andrew Nikiforuk points out that while the United States has at least gone some way to compensating both Native American and settler miners and associated workers or their families via the Radiation Exposure Compensation Act (RECA), Canada has never accepted any liability. This refusal to accept responsibility was despite most of the historical uranium mining being done under a nationalised industry. Nikiforuk goes on to critique the CDUT 2005 report for its overly narrow focus, which ignored both settler miner exposure and the lessons from the Navajo experience in the United States (Henningson 2006).

More powerful is Van Wyck's (2010) critique when he argues the upshot of the report is that while the community is seen as damaged from the history of colonialism, disjuncture with traditional practices, alcoholism, and depression, the mining industry is "officially exonerated." This seems confirmed by the fact the mining firm Alberta Star was granted a uranium mining licence near Port Radium in 2006, with at least sections of the Dine community giving consent for this (Henningson 2006). This shows, as Powell (2018) emphasises, that First Nations people are not homogeneous groups all evincing climate justice. Van Wyck (2010) also points out a strange circular irony in that "poverty of facts became the facts nonetheless" and in want of the withheld archive, and the report, via "bureaucratic logic," relies on the oral histories that positivist science holds as suspect. Despite this, "the testimony of their dead has proven insufficient. Once again, the living are passed over in silence" (Van Wyck 2010, pp. 171–172). More broadly, this is an example of coloniality where "stories passed down by indigenous elders cannot be considered true until they are validated by the empires that colonised them" (Marya & Patel 2021, p. 27).

Overall, this snapshot of Canadian nuclear history reveals a number of issues. Firstly, like all nuclear issues in the early period, there was a lack of understanding and health and safety measures in place. This prompted the United States to compensate workers, something Canada has yet to do and seems determined to block, which seems evident in the withholding of the archive of nationalised uranium operations. Secondly, we see in this history the concepts of nuclear colonialism, nuclear landscapes, communities, and peripheralisation played out. This involves the multiscalar relations of imperial and settler governments, Eldorado, and the Dine, matched by the relationality of economic, political, medical, and Dine relations to the land. Then there is the spatio-temporal element with the Highway of the Atom, a fitting concept that encompasses the spatial A to B and the temporal now to then, and so much more. Thirdly, the idea of accepting differing ontologies and epistemologies is rhetorically presented, but then in practice even the science itself is, in Nader's terms, somewhat unscientific. Then the tensions with Traditional Knowledge emerge, as it serves as a way of understanding the Dine's internal world, but not the objective external world, or the oral histories serve as poor substitutes for the withheld archive. This epistemic hierarchy will become more salient below in the sections that explore the Nuclear Waste Management Organisation (NWMO).

NWMO: Aims, Scope, and Assumptions

That Canada's NWMO is committed in principle to partnership and balancing of Traditional Indigenous Knowledge with scientific knowledge is stated explicitly in all major publications. For instance, the NWMO 2021 annual report is entitled: *Guided by Science. Grounded in Knowledge. Committed to Partnership*, and opens with a land acknowledgement that states the NWMO has worked with many First Nations groups since its inception and is working on a First Nations traditional territory (NWMO 2021a). This section will outline the broad aims, scope, and underlying assumptions of the NWMO, arguing that as an ideal it is something to be strived for as we must as societies deal with legacy nuclear waste in as just a way as possible. However, this endeavour is undermined by the conflations implied and assumptions embedded in these documents. These are, as Nader (2004) laid out, assuming growth is good and can/should continue indefinitely, conflating technological with social progress, the use of the god trick or objectivity to differentiate the Western eye from Indigenous Knowledge, and taking a very narrow temporal and sectoral view that essentialises incumbent technology as the only way.

The overall approach the NWMO has taken is known as Adaptive Phased Management (APM) and was chosen, in a similar way to the United Kingdom, after the failure of more technocratic approaches (Bell 2022; Cotton 2017). In outlining this approach, the NWMO presents this as a combination of a technical method and a management system. This technical expertise in this framework involves the building, a GDF in a suitable rock formation, and the waste itself in suitable and safe containers, while the management system involves a "phased and adaptive decision-making, supported by public engagement and continuous learning" (NWMO 2022, p. 7; NWMO 2021a). The public engagement and learning elements stipulate that a GDF will only go ahead with "informed and willing hosts," and this includes exploration of how the project could improve communal well-being (NWMO 2021a, p. 10). This project is "adaptive by design," which allowed it to adapt and continue work through the pandemic, and is guided by an ethical and social framework. This framework was developed from the CDUT 2005 report and round table (NWMO 2023a) and is as follows:

- Respect for life in all its forms, including minimisation of harm to human beings and other sentient creatures;
- Respect for future generations of human beings, other species, and the biosphere as a whole;
- Respect for peoples and cultures;
- Justice (across groups, regions, and generations);
- Fairness (to everyone affected); and
- Sensitivity to the differences of values and interpretation that different individuals and groups bring to the dialogue (NWMO 2023a).

The NWMO claims that it recognises the import of building good relations with the Indigenous people it works with, accepts injustices of the present and the past, and has thus "embarked on a Reconciliation journey" (NMWO 2021a, p. 20). This reconciliation includes not only activities such as mandatory staff reconciliation training, informal training, staff support systems, and an assessment tool, but also more general cross-cultural exchange of the settler technical staff with Indigenous cultural practitioners (NWMO 2020). This reconciliation has been depicted as a tree with the branches as temporal stages in the process, which exemplifies the claim that this institution is learning through its social engagement as it claims (NWMO 2022).

Of interest to this book is the idea of intergenerational democracy, so the branch that refers to 2021 and beyond is of special significance and will be briefly explored. The NWMO youth strategy has three broad goals: to raise youth awareness of the NWMO and APM project and aim to site in consenting community; to build youth understanding and assurance in the project and its processes; and to build youth capacity in becoming future decision-makers involved in the NWMO and GDF processes (NWMO 2021b). More specifically, this strategy involves student work placements in the NWMO and APM, Councils of Youth and Elders as community liaisons spurring new perspectives and contributions to the project, sponsorship and donations to youth well-being initiatives and educational programmes, and both early investment in education and skills and post-secondary institutional collaboration (NWMO 2021b). All this sounds very positive; however, one does wonder, as we will see in the next cases, why the government is not doing its job to look after and properly fund the next generation.

The NWMO (2021a) presents a timeline of what it sees as the overall process. This involves a number of stages, with 2023 being the year the specific GDF site was agreed upon. This timeline stretches back to the process begun 21 years earlier with the creation of the NWMO and the CDUT process, investigation, and report detailed in the previous section. It projects forward to 2040–2045, where it estimates the GDF will be completed and opened, and then the nuclear waste will begin to be placed in this repository. Significant points in this timeline include

- 2002–2007: developing the NWM plan;
- 2010–2023: identifying a site for the GDF;
- 2023: choosing a single site;
- 2024–2033: construction of the site; and
- 2040–2045: beginning operation of the site (NWMO 2021a).

The problematic assumptions embedded in the NWMO are not hard to find. The introduction, repeated verbatim in two documents of the NWMO (2021a, p. 8, 2022, p. 6), claims:

> Canada has been using nuclear energy as a reliable, low-carbon power source for our homes and businesses for nearly 60 years. Now, as worldwide energy demand grows and the need to address climate change intensifies, nuclear power has become an increasingly important part of the conversation.

A first problem here is the conflation of new nuclear builds with dealing with nuclear waste. This is understandable with much of the expertise and personnel taken from the nuclear industry and given the NWMO was established by the privatised energy companies with the Atomic Energy of Canada Limited (AECL), who also fund it (NWMO 2023b). However, it is argued in this book that this conflation is a mistake. If the aim is to get bipartisan support to deal with the legacy waste justly, this needs to gather support from those that support and oppose new nuclear builds. Also, the statement above presents nuclear power as "reliable" and "low-carbon" – both disputable depending on what we are comparing it against and who is doing this (Blowers 2020; Vettese & Pendergrass 2022). The NWMO would be in a much stronger position to deal fairly with all sections of the community – in line with the above ethical and social framework – if it drops this implicit support for nuclear power. The phrase "energy demand grows" is redolent of Nader's (2004; Hadjilambrinos 1990) taken-for-granted growth models and energo-power. This assumes an ever-increasing demand for energy, and many other things, rather than steady-state, sufficiency, and other ways of meeting human needs across space and time. This also implies a conflation of technological and social progress, which, as argued in previous chapters, is a hard sell considering neoliberalism's boomerang effect.

This boomerang effect involves the surveillance, oppression, and violence imposed over the last 500 years on colonised people, now coming home to roost upon the imperial core (Koram 2022; Davies et al. 2022). Thus, as peripheralisation reaches more lives and livelihoods, many will ask themselves what the point of new technology is if their lives become more precarious and difficult. Also, many colonised and First Nations people in settler states never shared in technological progress such as electrification even when the energy was extracted from and polluting their land (Paradies 2020; Powell 2018).

Finally, there is an implicit god trick that pervades these documents, seen not only in the passive voice that characterises the technical sections but also in its commitment to peer-review (NWMO 2023c). Peer-review processes are presented as some kind of fool proof and objective clearing house, when in reality it is just another human, positional process with various problematic and power-laden features (Erickson 2016). This recourse to peer-review also implicitly elevates science and expert knowledge above that of lay (Funtowicz & Ravetz 1993; Fischer 2000) and indigenous knowledge.

The Search for a GDF Site and Implementation

The NWMO claims to have narrowed its search for a GDF site to two locations, and as stated earlier, a decision on a final site was expected in 2023. These sites are both in Ontario – the first in the Wabigoon-Ignace area in the northwest and the second in the SON-South Bruce area in the south (NWMO 2022). In an earlier process regarding low-level waste (LLW) disposal, Hunhold (2002) describes how the various levels of Canadian nuclear community at national, state, and local areas begin with an open and democratic process. However, this democratic process is degraded by the narrow framing of locale to municipality, exclusive of surrounding areas, and more generally by governing interests seeing and using this narrow scope to control the process and achieve a solution. This resulted in failure with the proposed site of Deep River pulling out and ultimately an on-site solution being found that seemed fairer anyway as the waste stayed in the communities that benefited economically from the nuclear landscape (Hunhold 2002). The NWMO (2023d) appears to have learned this lesson and states that a site for the GDF will only go ahead with the site municipality, First Nations, Métis communities, *and* surrounding communities working together. However, as in Hunhold's (2002) case, the space between rhetoric and the actual process, or the discrepancy of promises and practice, is where we must search for and expose procedural injustices in any governing scheme. This is especially the case in something as polemical as nuclear waste.

From the CDUT process, one of the main arguments presented by First Nations and Métis communities was that they do at least some of their own research in their own communities. This resulted in two studies that can be used as reference points for the overall NWMO operation and the discussion below. Wanlin and Tiura's (2006) study explores how to develop a two-way communication between settlers and Aboriginal peoples. It presents various principles and methods that should be used, including not only treating Aboriginal communities with respect and acknowledgement of past injustice but also seeing these communities as distinct and not unified in need or character. It also recommends the NWMO do things differently, such as accepting there are differing ontologies related to land and the more-than-human. It argues that to build trust the NWMO must act with integrity and consistency and that it should respect refusal and leave the door open. It also recommends including both the youth of these communities and the traditional elder power structures, power structures that often differ in personel from the official democratic office holder. As argued in the previous chapter, this official office is often a structure problematically imposed on communities and often tied the property and heteropatriarchal dictates of the settler colonial society (Estes 2019; Powell 2018; Voyles 2015). Finally, it argues for aligning means with ends and an open approach that asks what people want to know about nuclear waste management, effectively allowing the community some control of the process (Wanlin & Tiura 2006).

The second study was conducted by First Nations students (Nevin and Piccin 2007) with some knowledge of nuclear waste and engineering. Their role was to go to various First Nations and Métis communities to talk about issues, provide information, and answer questions. This study shows that although many of these communities suffer from high rates of illiteracy, this should not be conflated with ignorance as the inquiries these students reported were the same or similar to those concerning experts such as Hunhold (2002) or Kuhn (1998). These questions included inquiries about NWMO's history and interaction with indigenous communities, what its mandate was, and information about the nuclear fuel cycle; inquiries about alternatives to the GDF and why these alternatives were rejected; inquiries about safety and security of a GDF and how natural disasters such as earthquakes might affect this; inquiries not only about the relations between nuclear power generation and the production of nuclear weapons but also the NWMO's position on nuclear power more generally, as there was the concern that a GDF site would encourage further expansion of nuclear power over renewable energy; and finally, enquires about the regional or surrounding communities' input, arguing that a regional approach was fairer (Nevin & Piccin 2007). Thus, taken together these studies commissioned by the NWMO, alongside the concepts of nuclear colonialism, nuclear landscape and communities, peripheralisation, and energopower, will be used to assess the democratic legitimacy of the NWMO's practice.

As stated earlier, two sites remain in contention for the GDF – the Wabigoon-Ignace area in the northwest and the SON-South Bruce area in the south of Ontario (NWMO 2022). Research was conducted in the second area by Bell (2021, 2022), who found that the promises of the NWMO did not match the practices and how these landed spatially. Firstly, Bell focuses on the epistemic injustices, involving the misrecognition of diverse First Nations and Métis communities as homogeneous (against the specific advice of Wanlin and Tiura above), and a downgrading indigenous knowledge in relation to Western *objective* science (Bell 2022). Secondly, Bell focuses on the broader procedural injustices of the NWMO process in Bruce, involving questionable claims of neutrality on the community liaison committees (CLCs), local elite capture, and silence interpreted as acceptance (Bell 2021).

In terms of epistemic injustices, Bell (2022) firstly points out that the deficit model of public understanding is often used in technoscientific decision-making. This involves a tension between lay/expert knowledge that has long characterised nuclear discourse (Wynne 1989) and the privileging of Western scientific expertise over other ontologies and epistemologies. In contrast to this Western technoscience discourse, Bell (2022) argues for a nuclear landscape lens that sees knowledge as place-based, relational, and situated, a position also argued for by theorists such as Haraway (2016). In this place-based and relational understanding, binaries such as expert/lay can be surpassed and a diverse constellation of knowledges and embedded expertise can be recognised. This might also allow recognition of technoscience's entanglement with coloniality, which largely necessitates technoscience's separation from and elevation above indigenous knowledge (Moore 2015; Bell 2022). The problems with the NWMO processes revolved around three aspects of epistemic injustice.

The first aspect was the homogenisation of First Nations and Métis people. This homogenisation of indigenous groups – again specifically warned against – involved the NWMO production of the Journey of Water series, an informational exercise presenting indigenous knowledge, scientific knowledge, and the likely impacts of a GDF on local water bodies. The problem with this exercise was it elided the 600 First Nations and 60 languages spoken and their heterogeneity. Instead, this exercise presented "institutional-level NWMO Indigenous representatives" that might be understood by municipalities as representing local indigenous perspectives and interests (Bell 2022, p. 16). A second aspect of this was the separation of dealing with First Nations and Métis communities with the process dealing with municipalities. Stanley (2009) argues that this is a discursive production of difference aimed at maintaining and reinforcing power relations and spatialised and racialised inequalities. In contrast, this can be seen as recognising First Nations as nations and so an act of recognition justice. However, the problem emerges when, as in this case, the information from the indigenous group does not reach the municipality, and this body makes its decision with little understanding of this indigenous perspective (Bell 2021, 2022). The third and more general epistemic injustice was mobilisation of the deficit model and the framing of the general public and groups within it who were critical of nuclear power as irrational/emotional. This deficit model was pushed by technoscience actors and the general public onto itself, with anti-nuclear discourse dismissed as "strong views" and "ignorant" by one retired nuclear engineer member of the CLC and "blah blah" and down to a lack of education by one public member of a CLC (Bell 2022, pp. 11–12). There was another example of a young man who asked a critical question and was inundated with "facts," presented from the Western technoscience discourse. This book believes we both create and discover the world and that separating these processes is difficult. This is why Habermas (1996) argues science must address the prior and higher-order philosophical questions of *why* and *should we* before the scientific questions of *what* and *how.* In dealing with nuclear waste, this requires a more sensitive and balanced form of communication. Bell (2021) details numerous people who felt uncomfortable with the technical information and how this was communicated, which can be connected to the tension between simplifying information to communicate essential evidence/data and communicating in a disrespectful and condescending way. Finally, the CLCs were presented as neutral (Haraway's god trick) when at least the Bruce CLC was clearly far from this, with anti-nuclear voices excluded or disparaged in this venue. This contrasts to the Northern Ontario experience, which was much more diverse and included critical voices, and is thus an example of how a nuclear landscape influences the knowledges produced, mobilised, and legitimated (Bell 2022).

Procedurally, the NWMO practices appear to have allowed both a false consensus to emerge and local elite capture. As stated earlier, the idea that CLC was neutral was problematic but here involved a remarkable conflation that is very important to the arguments of this book. One member of the CLC explained (Bell 2021, p. 171):

> I think everybody that sits at that table is very pro-process and pro-solution. I don't think that they came to the table as anti-process, anti-nuclear.

These conflations of nuclear waste siting with nuclear power generation are problematic. This is because something as unpopular as nuclear waste siting (Kuhn 1998) demands the inclusion of whole communities, not just those that support nuclear power. Thus, the idea that those who oppose nuclear power are anti-process of dealing with nuclear waste is wrong and unhelpful. Other community members were fairly dismissive of this pro-process neutrality, with one pointing out that the CLC in Bruce was, with one exception, composed entirely of former employees of the nuclear industry (Bell 2021). Finally, there were injustices reflecting Hunhold's (2002) concern about democratic localism, and broader environmental justice concerns about unchecked decentralisation and local elite power dynamics (Walker 2009; Lemos & Agrawal 2006). This concerns the local mayor of Huron-Kinloss chairing a CLC, who was also a real estate broker, and another leader of a CLC being a farmer receiving direct payments from the nuclear industry. Whether this entailed conflicting interests was not as important as the perception that it might, and this was evident in at least one person's self-censorship due to fear of the mayor's ability to harm their business. More broadly, this mayor explained that his role was to "avoid conflict" via the use of "procedural bylaws," which can be seen as imposing consensus and as such an anti-democratic move (Bell 2021, p. 173).

Conclusion

Overall, the promises of the NWMO are not being met by it spatial practices, which, to be fair to this organisation, is arguably a more general and endemic feature of big institutions and how they operate. Firstly, Canada must deal properly with the colonial history, and an essential epistemic justice first step of this would surely be opening the archive. Transparency has always been an issue of the nuclear, as we will see in the later chapters, but it is also a key part of the environmental justice of process and recognition. Also, the problems Bell (2021, 2022) describes in the Bruce processes seem to undermine some of the basic ethical and social principles the NWMO (2023a) espouses. These problems also undermine Wanlin and Tiura's (2006) notion of two-way communication, while also eliding the legitimate concerns of indigenous communities reported by Nevin and Piccin (2007). For example, in the Bruce case, how can imposing a false consensus or group thinking of mainly ex-nuclear employees and their supporters be an example of "Respect . . . Justice . . . Fairness . . . for peoples and cultures . . . and generations," or "[S]ensitivity to the differences of values and interpretations that different individuals and groups bring to the dialogue"?

There is also ontologically and epistemologically an implied hierarchy of worldview and knowledge. This in many ways is an aspect of mainstream science that continues, in Nader's terms, to frame itself as autonomous and culture free. This is perhaps partly due to the reluctance of many mainstream scientists to accept science's historic role in categorising, mapping, quantifying, and, more generally, developing the technical knowledge that enabled the successive phases of capitalist colonisation of land and people (Moore 2015; Rushcoff 2022; Harris 2023). The fact that the NWMO board is made up of five senior nuclear engineers/managers and four senior business/accountants

(NWMO 2023b) is a modern example of the fusion of capitalism and science that would likely be replicated across a neoliberal and privatised world of energy. This fusion is documented memorably by Harris (2023; Gilmore 2023) in his history of Palo Alto, Stanford University, and the development of the capitalist frontier of California. This is a history of pseudo race science and eugenics, ethnoclass oppression, and the growth of the military industrial complex (Harris 2023). If the process of rationalisation of land and resources is slowing, it is not for want of various techno-scientists offering new frontiers to colonise, be that geoengineering (Sovacool et al. 2023), biopiracy (Moore 2015; Shiva 2007), or even space (Pien 2022). Rather it is the lack of required profits for the investors, social exhaustion due to the ever-widening peripheralisation creeping up the ethnoclass structure, and an active, disturbed Earth system that will not and cannot be managed (Christophers 2023; Fraser 2021; Moore 2015; Dryzek & Pickering 2019).

On a more positive note, Canada and the NWMO could make this a much more just process with some relatively easy steps: firstly, open the archive and set up a compensation scheme similar to the United States' but learning from its short comings; secondly, align the Bruce case with the other case which seems much more open and fair; thirdly, stop conflating nuclear power with waste solutions as the latter requires the whole community buy-in not just those that support nuclear power; and finally, the tensions between scientific knowledge and indigenous knowledges are beyond one organisation or one state; therefore, this tension will be considered in the conclusion.

7 The World's First GDF: Finland

Finland is set to be the frontrunner in developing a GDF, and its NWMO, Posiva, aims to cash in on this and share its learning internationally (Posiva 2021). However, some lessons and learnings travel well, others not so well. This is one of the key issues this book has with scientistic and reductive approaches. While the technocratic and technoscience expertise underpinning the safety case of nuclear waste management in Finland may travel well, the context and history and the way the nuclear industry and NWMO are placed in and interact with society are particular to this country. Nevertheless, there remain some parallels such as elements of nuclear landscapes/communities and peripheralisation.

Therefore, this chapter will, firstly, outline the Finnish nuclear context and history. Secondly, it will examine the aims, scope, and assumptions of the two main institutions involved in Finnish nuclear waste management, Posiva – the private company overseeing the GDF project, and the radiation and nuclear safety regulator (STUK), and more briefly, the two nuclear power operators. Finally, it will look at how these institutions have implemented this process. Overall, the argument will be that much of the learning from the Finnish case would not apply to other countries unless they had the same or similar geology and trust in institutions. However, there are lessons to learn from this case, including certain democratic lacunae, how benefit systems work and do not, and balancing procedural and distributive justice across time and space.

Context and History

Finland has two operating nuclear sites, one in Loviisa with two pressurised water reactor (PWR) units and another in Olkiluoto with two boiling water reactors (BWRs), and a third PWR that opened in 2021 (STUK 2022). There is a sixth site that is due to open but has been delayed. Overall, this nuclear power supplies approximately a third of Finland's electricity (Lehtonen et al. 2021). Thus, Finland is attempting to expand, if modestly, its nuclear energy at the same time as dealing with its spent fuel. Blowers (2020) argues this "creates a tension," which as we will see later, is a tension felt within the Finnish nuclear

DOI: 10.4324/9781003422686-10

communities. Finland does not mine its own uranium and so does not have the problematic socio-environmental history and present of this mining industry. However, like any other importer, it does farm out or externalise these problems, which raises two linked spatio-temporal justice questions.

Firstly, if a supposedly low-carbon technology relies on mining for its resource, how does it ensure that this mining is done in a socio-environmentally benign way? Currently, the multinational mining firms, largely based in the minority world, operate in the majority world, where there is both a lack of regulatory capacity and, in many cases, the corruption and collusion that accompany what is known as the resource curse (Van de Graaf & Sovacool 2020). At the same time, in the minority world, over 40 years of neoliberal erosion and capture of regulation have taken its toll (Koram 2022). A deeper problem lies in the nature of uranium itself, which, in Hajer's (1995) terms, has been organised out of the category of what Hecht (2012) labels nuclearity – which in short refers to the technopolitical processes that determine which objects are nuclear and which are not. Thus, despite the harmful effects of radiation being known at least since 1879, health researchers and activists working with African uranium miners, a significant source of world uranium (WNA 2023b), have had to continuously reassert and reprove this causal link, with the industry resisting. In Hecht's (2012) reading, uranium mining and trading and its political economy take on a decidedly post/neocolonial aspect, with most of the advanced part of the nuclear cycle being a "first world" concern with attendant safety protocols, while the mining for the fuel, at least in Africa, takes on a backward and historical "third world" character and therefore lacking the said safety protocols. More prosaically, uranium is not a renewable resource but a stock similar to fossil fuels, and although it is relatively abundant now, this situation could change quite quickly if the nuclear industry expands as rapidly as some advocates wish (Hansen 2009).

Mining as a process is one based on diminishing returns with a number of effects. Firstly, ore grade falls with the consequence that the average carbon equivalent (mining, building, etc.) intensity would likely rise from 34 gCO_2/kWh based on current 0.15 grade purity to as high as 60 gCO_2/kWh (Norgate et al. 2014; Frank 2022). An earlier estimation puts nuclear power equivalent carbon intensity between as wide a range as 1.4 and 288, with an average of 66 gCO_2/kWh (Sovacool 2008), so the above figure should be treated with caution, but the overall point is that this figure will only increase with time. Under capitalism, even relatively abundant resources such as oil or food are routinely manipulated into imposed scarcity to inflate the price (Mitchell 2013; Harris 2023; Sahlins 2017). So, one can imagine that a combination of market manipulation plus real scarcity, depending on how fast the nuclear industry expands, will push the price of nuclear energy upwards. Thus, temporally nuclear energy is in this sense pushing the problem of long-term solutions to the CEC into the future as at some point uranium will become too scarce to be a viable fuel. This is where nuclear advocates will point towards various innovations and future tech solutions, like fast breeder technologies that seem beset with issues (Vettese & Pendergrass 2022), or nuclear fusion, the classic silver bullet. A problem with both these innovations is that they are not ready and beset with problems despite the massive amounts of funds ($100 billion around the world on fast breeders) sunk into them (von Hippel 2021). Thus, these solutions can be seen as gambling with people's, flora, and fauna lives and ecosystem functioning.

Finland's spent nuclear fuel (SNF) is kept in interim storage at the two Olkiluoto and Loviisa sites, and as of 2021, there was approximately 2,400 tonnes of the SNF cooling in preparation for final disposal (Posiva 2021). The search for a site went through a similar process to the Canadian and British states, with the failure of a technocratic, top-down process forcing a more deliberative approach (Kojo et al. 2010). Legislation banned the 1980's practice of exporting waste (to the USSR), leading to the then state-owned operator IVO (now Fortum, 51% owned by the state) and the private operator TVO to establish the NWMO Posiva in 1995. TVO had been looking for a GDF site since the 1980s, but local opposition had forced it to pay greater attention to local stakeholder views (Kojo 2009). This deliberative turn deepened in 1997–1999 as Posiva arranged "an exceptionally long and participatory process" of Environmental Impact Assessment (EIA), a process made mandatory shortly afterwards (Lehtonen et al. 2021, p. 135). The municipal council of Eurajoki gave its approval for the GDF in 2000 (Kojo 2009), and this has been in development since then and is set to open imminently as the world's first.

Overall, this process must be put into the context of a society with high levels of trust in both the government and the industry. Most recently, on the question of how satisfied the Finnish were with their democracy, 83% were satisfied, 16% unsatisfied, and 1% unsure, making Finland the fourth most content in the EU on this metric (Standard Eurobarometer 2022, p. 24). Regarding trust in government, Finland again does well, with 56% claiming they tended to trust their government to 38% tending not to trust their government, which places them in fifth EU-wide and compares to the United Kingdom's 62% tending not to trust the government and only 34% who say they do tend to trust their government. Trust in local public authorities rises in both countries with 72%–23% in Finland and 52%–41% in the United Kingdom (Standard Eurobarometer 2017, pp. 42–45). This alongside a proportional representation electoral system that can engender better representation and be less dysfunctional than zero-sum first-past-the-post systems places Finland in a fairly stable political context. This stability contrasts to a general crisis of legitimacy and trust in liberal democracies (Quilter-Pinner et al. 2021).

Aims, Scope, and Assumptions of NWMOs in Finland

There are three main interacting groups/institutions dealing with Finland's nuclear waste. These are the government's Radiation and Nuclear Safety

Authority (STUK), which oversees all nuclear operators and their activities; the nuclear operators Teollisuuden Voima Oyj (TVO) and Fortum Power and Heat Oy (Fortum) responsible for managing interim SNF and low-level waste (LLW) and intermediate-level waste (ILW) disposal; and Posiva, owned by TVO and Fortum and responsible for the GDF and final disposal of the high-level waste (HLW)/SNF (Posiva 2021). These institutions will be examined in turn before exploring the progress on the GDF Posiva is developing as the most advanced in the world today. Again, as with the Canadian NWMO, it is important not only to examine what they say about their aims, objectives, and scope but also to interrogate the implicit assumptions.

STUK

As the overall government regulator of the nuclear industry, STUK has a broad scope covering the safety of this technology. This includes aligning and informing Finnish nuclear safety with regional and international standards and knowledge, updating general and site-specific regulations for nuclear sites existing and planned, and conducting probabilistic risk assessments (STUK 2022). In terms of aims, it published a strategy in 2018 covering the period 2018–2022, and this included a number of strategic goals. STUK's (2018a, p. 2) overarching aim is to "ensure radiation safety of people [including future generations] and environment," and this aim is guided by four values:

- The expertise of STUK actors and their up-to-date scientific knowledge which enable them to understand complex entities;
- The openness and transparency, which include accepting new information and ideas while also providing clear information about STUK and nuclear safety;
- The courage to try things differently and explore difficult issues, while assuming responsibility and learning from mistakes; and
- A co-operative mentality based on "trust, interaction and respect for everyone's work" (STUK 2018a, p. 4).

From this broad aim and these four values, STUK (2018a) produced nine strategic targets grouped into three categories: targets related to STUK's resources, targets related to STUK's effectiveness, and societal targets. The three targets (STUK 2018a) related to resources include enabling the happiest civil servants in the world, understanding complex entities, and being cost-aware. The first of these targets is framed in human resources speak and instrumental terms, as happiest equals best and is facilitated by the correct forms of leadership, administration, and organisation, allowing workers to "concentrate on essential tasks." The capacity to understand complex entities combines cognition sufficient in both depth and comprehensiveness, while overseeing operations in a "risk-informed manner," communicating these risks effectively and successfully, and mitigating these risks. Cost-aware operations echo the usual neoliberal refrain of do more for less (Peck 2010; Harvey 2007), as STUK now aims at "lightening the cost structure" and is more cost-aware than before (STUK 2018a, p. 8).

The three targets related to STUK effectiveness include risk-informed oversight, flexible and efficient working practices, and effective radiation safety research. STUK's (2018a, 2018b) risk-informed oversight is based on overall safety of operation and operators in proportion of risks of the plant or operators, which includes not overseeing radiation risks deemed "minor." Flexible and efficient working practices refer to results and effectiveness "regardless of time and place and ensuring co-operation and a feeling of togetherness." This suggests a progressive attitude to working methods that in contrast were only realised, for specific groups, in the United Kingdom under the pandemic lockdowns and which have been contested by certain interests ever since (Partridge 2022). Effective nuclear safety research involves collaboration with research institutes and universities both domestically and internationally and making sure that financing is available for this work (STUK 2018a, p. 10, 2018b).

Finally, the three societal targets involve emphasis of the safety responsibility of the operators, people's understanding of the risks of radiation, and society's resilience to disturbances. STUK (2018a) argues safety is not the result of inspections but a culture of safety within an operator, and thus STUK's role is to positively nurture and coach this safety culture and those responsible for it. This involves more STUK involvement when risks are deemed as high and less when they are deemed low or when the operator can prove that the "safety level of their operations is good." Providing and communicating risk to the public involve sharing the "correct and easy-to-understand radiation safety" that will enable the public to understand these risks and act "correctly without unnecessary fear," regardless of disinformation or information overload. This rather problematic framing of risk as objectively definable will be returned to below when considering the assumptions. Society's resilience to disturbances, a somewhat disturbing notion in itself, regards the "overall safety model" and preparations for "new threats" ahead of time and the goal to "secure society's ability to function," via clear systems and responsibilities, requisite resources for an unspecified event (STUK 2018a, p. 12). Again, this rather vague but disturbing target will be returned to when considering the shared assumptions of Finland's NWMOs.

TVO and Fortum

TVO and Fortum privately and partially publicly operate the nuclear power plants (NPPs) in Finland and are responsible for operational and waste safety. Operationally, this will be under the regulation of STUK, with more oversight for more risky operations, but mitigated if the operator proves a good level of safety or if the process is deemed low-risk. As a private company, TVO's overarching aim will be the fiduciary responsibility to extract profits for shareholders, while Fortum's 51% ownership by the government may at least mitigate this fiduciary priority. Corporate fiduciary responsibility, some argue, can be balanced by environmental and social governance (ESG) aims and objectives. However, globally evidence suggests ESG is more about building resilience and sustainability of large extractive corporations and economic institutions, rather than social and environmental sustainability (Gabor 2022; Buller & Braun 2021; Dafermos et al. 2021). In terms of waste, TVO and Fortum are responsible for the interim storage of SNF for its cooling period before it is transferred to the GDF and the disposal of LLW and ILW. This disposal of LLW and ILW is conducted in or near the existing NPPs at Olkiluoto and Loviisa in underground facilities (STUK 2022).

Posiva

Posiva is a company owned and funded by TVO and Fortum with the aim to develop the GDF, which also monitors the ongoing safety of LLW and ILW processes. One of the major aims and responsibilities of Posiva is to establish the long-term safety case of the GDF, and this involves four areas:

- Arguing the SNF and contained radionuclides "will likely remain" in the Olkiluoto bedrock and copper canisters for "more than a million years";
- Developing a GDF based on natural and engineered barriers that will not be compromised by geological or climatic events of processes, with performance assessments over the different required time scales suggesting possible release of radioactivity in the future will remain below STUK's dose guidelines;
- Developing a biosphere analysis investigating radiation doses of the most exposed populations, involving a "simplified dose calculation" across a million years that shows that even under pessimistic development scenarios, these doses will be fractions of natural radiation doses; and
- Showing that ILW and LLW created in the process of dealing with SNF can be reposited at this site at a depth of 180 metres (Posiva 2021, pp. 5–6).

Posiva is a private company and so aims to profit like any other. As such, it reports remaining profitable and growing throughout 2021, and it is a company that offers "ready-made solution models and service systems" for NWMOs internationally and aims to take advantage of its early adopter status by selling services to various other states including the United Kingdom (Posiva 2021, p. 13). However, this early adopter status is just as easily framed as first-mover disadvantage, whereby the first to move makes the costly social, environmental, and financial errors that in a megaproject such as this are hard to reverse (Cotton 2017). Simultaneously, those watching can learn from this and avoid these problems. This is not an argument to do nothing or to wait until somebody else goes first but rather to show both sides of this first-mover dynamic which Posiva does not.

Finally, Posiva is responsible for the construction and encapsulation of the GDF in its entirety for the required time. This involves both engineered and natural barriers, including an inner and outer canister, buffer, and bentonite backfill in the deposition tunnel, and all this within 400–500 metres of bedrock. This can be seen as a system that has multiple layers of redundancy and so, according to Posiva, will, even under more pessimistic scenarios of future social and natural events and processes, not cause socio-environmental harm (Posiva 2023).

The GDF is placed in the crystalline bedrock that makes up much of Finnish geology at a depth of approximately 430 metres. The final facility will include the above-ground encapsulation/processing facilities and the access tunnels leading to the SNF storage areas. Overall, Posiva makes the bold claim to have "solved final disposal in a safe manner" – a precondition of nuclear power operation into the future (Posiva 2023). However, Posiva also leaves a "back door open to technological progress" in finding a better solution, suggesting the "solved" above is provisional and there is some implicit recognition of first-mover disadvantage. However, this "back door" must be used in the near term (next 100 years), as after this the GDF will be closed permanently and retrieval will be more challenging (Posiva 2023).

Shared Assumptions

The implicit assumptions of these institutions are not difficult to find and can be grouped into three. The first is seen in the statement above that Posiva has found a "safe" disposal solution upon which the future of Finland nuclear industry depends. That Posiva is owned and funded by the very industry that has a direct interest in the safety case's credibility is surely a conflict of interest. A second and perhaps more serious problem is the notion purveyed of the "public's correct understanding of risk" alongside being "prepared for and resilient to disturbances" (STUK 2018a, p. 2). The idea of *correct* understanding of risk is scientistic, to say the least, and undermines the idea that STUK, Posiva, and the nuclear industry have learned anything about risk perception, with the implication that formal risk analysis and the simple use of quantitative methods and then communicating this are all we need to fill the deficit (this model again) of public understanding (Wynne 2006). It also seems to occlude how nuclear risks in particular are socially constructed by both nuclear advocates and opponents (Cotton 2017). We have seen these framing effects with the nuclear industry adept at moving from discourses arguing we need nuclear power to support economic growth (Nader 2004; Hadjilambrinos 1990) to current securitisation discourses on how it will save us from the CEC (Cotton 2020; Blowers 2020). Then even in this short passage and repeated in point nine of STUK's strategic goals (STUK 2018a) is this strange and foreboding idea that society must be prepared/resilient to *disturbances*. Is this part of the *correct* understanding of risk? That society and presumably those closest to a facility must be prepared for some nuclear event even though this probably won't happen? This is contradictory to say the least.

A third problematic assumption is with the overall temporal aspect of the solution. A million years is a long time and not one we can model socio-environmental events or processes with accuracy for (Ramana 2020). There is an irony here in the inability of many scientists, most politicians, the mainstream press, and much of the public to see past a capitalist system that is at best 500 years old (Moore 2015). However, simultaneously we are trying to push one of the manifold problems of this system a million years into the future. This is comparable and connected to the hubris of geoengineers aiming to transform the ecosphere rather than a 500-year-old social system (Rushcoff 2022; Vettese & Pendergrass 2022). Communicating the dangers of this site over time is also problematic. Skrimshire (2018) considers the merits of different types of messages and communication but thinks that all seem to assume failure of communication. He then ruminates on what form a message should take, settling on the collective confession. However, Skrimshire asks is this a confession of something too late, with the impacts historically determined, or aimed at contemporaries, or somewhere between? This ambivalence in the question strikes a theme that connects this chapter to the last and the next. This is the question of the fundamental purpose of a GDF – is it to safely bury a mistaken technological choice, or is it to extend and expand the use of a problematic technology? To be fair, these temporal problems and questions are going to face any GDF, and so Finland is not alone with these issues.

Implementation of the Most Advanced GDF in the World

It can be stated from the beginning of this section that Finland is held up as a good example of nuclear policy including waste because it is, but also internationally this is a low bar to exceed. This suggests, as Winner (1980) argues, nuclear technology is problematic in terms of democracy and, that other social good, peace. Nuclear's link to the security, military industrial complex is indisputable, and as this is written, with the war in the Ukraine, and the role of NATO, the United Kingdom, and the United States in prolonging this conflict (Watkins 2022), nuclear war is an ominous shadow cast across the world again. This must not be seen as an endorsement of Russian aggression, just a plea for the only real safe and just option – a ceasefire and peace process. Back to Finland, as we will see, a *more just* and *democratic* process does not equal a process that is acceptable in terms of environmental/energy justice as fair process, distribution of risks and reward, and unbiased recognition of social groups (Cotton 2020; Sovacool et al. 2017). This is perhaps the core of the problem. Nuclear is a technology that is socially divisive and has very broad spatial and temporal footprints. These broad socio-environmental effects are true, even assuming no serious nuclear events occur soon, which is arguably not the safest assumption (Wheatley et al. 2017). Thus, the nuclear has difficultly being just.

This section will examine the implementation of Finnish GDF, examining this from a number of perspectives. Firstly, it will briefly outline the stages and processes by which the decision was made for the locale of the GDF, including what community benefits were agreed and how these were interpreted. Secondly, it will briefly examine how the media frames this issue, including a consideration of where Fins in the GDF area seem to get their information from. Thirdly, it will explore the procedural and distributional intra-/intergenerational justice issues including what the local community thinks about these issues. The overall argument will be that although this process looks fairer than in many other states, there is still evidence of Blowers' (2020) peripheralisation and the divided communities that we see in many other nuclear landscapes. One way this book has suggested overcoming, or at least mitigating, these problems is to explicitly separate nuclear waste processes from the nuclear industry and its future ambitions. Failure to do so risks exacerbation of the social tensions, which clearly exist even in a highly trusting and democratically content society such as Finland (Grönlund et al. 2010).

Finland's Search for a GDF

Eurajoki has long been a nuclear community with the first NPP constructed in the 1970s. However, despite being a community that housed an NPP and interim SNF and LLW and ILW, the municipality had a written understanding with TVO that final disposal of nuclear waste would never happen in Eurajoki, with this in fact being a condition of the original NPP (Kojo et al. 2010; Kojo 2009). By the 1990s, Posiva and STUK had narrowed host sites down to four locales (including Eurajoki chosen by TVO, thus undermining its written understanding), concluding they were as sound as each other in terms of geology and ecosystem; thus, it became much more a question of local acceptance (and who the local is in this formulation). This was because the Finnish Nuclear Energy Act of 1987 gave the host municipality the right to veto the GDF (Kari et al. 2010). This resulted in these four communities and other test/control communities becoming subject to multiple surveys, for this veto meant the nuclear industry and Posiva needed half the local councillors to agree, and under PR, this amounts to around half the local residents (Kari et al. 2010). The municipality of Eurajoki during the 1990s changed position, accepting TVO's plans for a new NPP and Posiva's siting the repository there (Kojo et al. 2010). The question is why the municipality changed its position, with Kojo (2009) suggesting that this was about a crisis of funding in the local municipality.

Describing this as "twenty years of arm-wrestling in Eurajoki," Kojo (2009) offers a description of how the decision turned. TVO, as suggested earlier, had included Eurajoki as a site despite their written assurance against this and the council's statement rejecting a GDF. Things began to change in the early 1990s with the change in the balance of power after a local election, the local chapter of the National Coalition Party proposing that the negative statement about disposal be removed, and the local council starting to cooperate with TVO. In December 1993, this proposal was blocked by one vote, but by December 1994 the National Coalition Party had a proposal to remove the sentence, rejecting the GDF accepted by 15 votes to 10. Thus, the municipality's objection to the GDF was neutralised. The underlying reason, Kojo (2009) suggests, was the planning proposal submitted by TVO for the GDF in a neighbouring municipality, which implied that approximately €1.17 million in real estate tax revenue might be lost. This was in the broader financial context of TVO between 1984 and 1990 paying a presumptive tax of €18.5 million, around a third of the municipality's tax income. This tax was abolished in 1991 and replaced by the real estate tax. Local councillors set up a working group with local businesses and went about an analysis of opportunities and threats. These opportunities and threats were mainly financially framed, with fears of the GDF going to another municipality seen as undermining the tax revenues and autonomy of the region. Accepting the GDF was also seen as helping establish the newly proposed NPP, which, it was expected, might bring in €2.5–3.4 million yearly. Thus, Kojo (2009, p. 179) argues these local politicians discounted environmental threats or local image risk associated with the GDF, instead choosing to "retain Eurajoki's position as an oasis of the nuclear energy industry." If this is true, this suggests an aspect of economic marginality in Blowers (2020) framework of peripheralised nuclear communities.

Returning to the question of who is the local in this local acceptance, there are questions about how democratic this siting decision was. The local here is firmly the politicians, and they wanted to keep it this way (Kari et al. 2010). Also, there was no funding for independent review or deliberative forums, as there was in Sweden (Vilhunen et al. 2022). This lack of funding and deliberation places lay communities in a poor position to deal with highly technical knowledge and the one deliberative process in this siting was the aforementioned EIA, itself a highly complicated and technical process (Vilhunen et al. 2022; Lehtonen & Kojo 2019). Thus, while Finland with its proportional representation system is likely more democratically responsive than winner takes all systems, representative democracy is arguably not enough to decide such issues of siting nuclear waste. This is because this places undue demands and risks from the collective onto a small group, and these asymmetrical distributions of illth and risk cannot be fairly conducted in representative democracy (Cotton 2017). Thus, we can add an element of democratic marginally to the issues with this process (Blowers 2020).

Finally, the community benefits seem modest at best and were again decided undemocratically behind closed doors. Lehtonen and Kojo (2019) frame benefits as problematic in that they can often seem like bribes or attempts to buy the local population, an issue others discuss (Blowers 2020; Cotton 2017). As a framework of understanding community benefits, these can be defined as *mitigation* or the reduction of actual or perceived harm via training, institution building, housing, or health services; *compensation* or payments for taking on the societal risk in the form of direct payments, contingency funds, or property value guarantees; and *incentives* or local empowerment, deliberative processes, or liaison committees (Lehtonen & Kojo 2019).

The EIA can in this framework be seen as both mitigation and incentive. It serves to mitigate risks by anticipation and/or remediation and as an incentive by the participatory element, but again this was undermined by the lack of funding for community events and deliberation. A further incentive can be seen in the municipality's option to opt in and out, although, as Lehtonen and Kojo (2019) argue, this can only be semi opt in, as TVO and any other NWMO decide the short list prior to any community opting in. Compensation was the aforementioned real estate tax, which in 2018 was 3.1% for nuclear installations compared to 0.92–2% for other industries. This incentive was likely a major motivator for the local politicians, as Kojo (2009) claims, and was accompanied by a complicated series of loans and property deals all negotiated behind closed doors, again suggesting questionable democratic processes. Lehtonen and Kojo (2019) state that in monetary terms, these benefits are a touch modest, in comparison to 2 billion SEK negotiated in Sweden a few years later. They go on to conclude that acceptance in the Finnish case could be framed as "somewhat of a pragmatic – even fatalistic – reaction in the face of the inevitable" or a "bought voluntarism" (Lehtonen & Kojo 2019, p. 196).

Media Representations and Consumption

This book discussed how the mainstream media works in Chapter four. To recap briefly and place this in the context of media communication of nuclear issues, there are a number of theoretical lenses generally used to examine this communication. This includes framing, discussed in Chapter four; focus on the event (usually a nuclear disaster such as Chernobyl); the idea of news factors or what is deemed newsworthy; and journalistic tone and bias (Kristiansen 2017). Clearly these aspects are not separate, and one could argue that framing largely covers all these factors, for this includes the selection of issues (news factors), in a certain light (bias) with a moral evaluation and solutions offered that usually benefit the powerful and status quo (Zabern & Tulloch 2020; Herman & Chomsky 2002). In terms of how the media frame all things nuclear, this is different over space and time, and this is based on a variety of issues. Firstly (Kristiansen 2017), over time media portrayals have moved from the techno-optimist era of the 1950s, through the 1960s where bomb testing and protests emerge, through the 1970s where this protest movement grew, and then the 1980s with two nuclear disasters leading to more critical coverage. The United States generally has maintained a positive, if cautious, media coverage, while in Europe this is much more varied based on nuclear events, how much nuclear energy is used in the state, the strength of the anti-nuclear movement, and when referenda or permitting of sites come up (Kristiansen 2017). Considering these issues, this section will examine how Finnish media covers nuclear issues and specifically nuclear waste, while also considering what media and information people are likely accessing.

On the whole, the mainstream Finnish media seems to treat the nuclear industry in a fairly friendly way. This can be seen in two cross-country comparisons of media coverage: one with France and the other with Sweden. Lehtonen et al. (2021) compare media framing of nuclear waste policy in Finland and France, focusing on reporting from the leading newspapers in each country. In Finland, this is *Helsingin Sanomat* (HS), the only national daily newspaper and the second most trusted news vendor after the public broadcaster, and is read by 17% of the population, with 27% reading it online. It declares itself independent (which newspaper does not?) but holds significant power, in international comparison, in its central monopoly of daily print news and the resulting ability to shape public discourse. From this paper, Lehtonen et al. (2021) extract six frames, with definitions of problem/situation, diagnosis/causes, moral judgement, and solution offered, with three of the five characterised by confidence in nuclear waste issues, while two are more ambivalent and one concerns arguments within the nuclear industry.

These six frames include:

- confidence in the project governance;
- confidence in the integrity of the GDF concept and Posiva's conscientiousness;
- confidence/scepticism in the ability to communicate with future generations;
- confidence/scepticism in nuclear power as a part of a low-carbon future;
- trust breaking down between nuclear waste stakeholders; and
- confidence in Finland as a GDF forerunner (Lehtonen et al. 2021).

The first frame of confidence in the project is prominent and present throughout the authors' data. It focuses on how the project is on schedule, how Posiva's diligent work means no surprises, and a realistic timetable. Morally, the frame sees this as good work that leaves time for the unexpected or complications, and as this frame is so positive, no solutions or remedies are necessary. The second frame is similar but goes a bit deeper into the GDF detail with mention of copper canister corrosion or an ice age as diagnosed problems, but the Finnish bedrock as the most stable in the world, meaning again that no further solution is needed. The next two frames offer more criticality with, firstly, the problem of communicating with future generations and the long time frames seen as the cause of doubt about how or even if to try to send a message. The fourth frame asks whether nuclear energy is environmentally friendly with this diagnosed as positive in that it is low-carbon but negatively in the waste it produces. For both frames, the moral judgement is ambivalent with the recognition that communication will be difficult over such a long time that it might be better to not send a message, while as an energy source whether this is irresponsible or the only option is the moral ambiguity. The remedy for time is to leave it to the next generation to worry about, while as an energy source it is decided that nuclear power and renewables can live happily together. The problems of trust between companies involves the disagreements between companies regarding the new NPP company Fennovoima's waste. The problem is that the GDF was seen originally as a national solution, but Posiva being owned by the other two nuclear operators seems to look after their concerns. Morally, both sides are seen as at fault, and the solution is all sides should cooperate and Posiva should share its know-how. Finally, the frame of confidence in Finland's role as GDF forerunner sees this part of a wider process with other leading countries such as Sweden or France, and down to good parliamentary processes and strong local support. The moral judgement of this focuses on how Finland will be the world's leader in this field, and this leads to the solution of disseminating information about the qualities of the Finnish project. Overall, the French media was seen as much more critical than the Finnish media, to which the authors offer two interpretations. The first is that this is unwarranted trust and a failure to present the bad with the good, and thus a lack of vigilance. The second is that Nordic democracies are cooperative and trusting and that media focusing on conflicts could erode this foundation of trust as well as undermining any emergence of a "pride effect" in the GDF area (Lehtonen et al. 2021).

The comparison between Sweden and Finland follows a similar pattern. Kojo et al. (2020) conducted their longitudinal study of media nuclear discourse focusing on two leading newspapers in Finland and Sweden. They found that on the whole Swedish newspapers (around 55% of articles) take a much more negative tone than the Finnish (around 27% of articles). Also, this positive discourse in Finland is interestingly upheld over the period of the Fukushima disaster. Kojo et al. (2020) conclude that Sweden has a more balanced approach including both sides of the debate, while Finnish media culture is more closed and favours core insiders of the nuclear industry and politicians. This they argue is a problem for Finland, in that a more open discussion can lead to higher public awareness and robust political decision-making, while they also posit that this "lapdog" approach by the Finnish media could partly explain why the GDF process went so smoothly (Kojo et al. 2020). This would correspond to earlier research on discourses surrounding energy security and nuclear new builds, which claimed Finland lacks a "critical public sphere" independent of government and the media is "largely pro-nuclear" with a "strong faith in techno-economic 'progress'" (Teräväinen et al. 2011, p. 3441). Thus, these papers tend to suggest a dynamic similar to the argument presented in Chapter four concerning the media in the United Kingdom. This is a media model that tends to manufacture consent to support the status quo, rather than reporting polemical issues in a more balanced way.

A final consideration of this section is where people in the GDF area tend to get their news. Kojo et al. (2010) look into the role of community and their views on the GDF plan in Finland, and one of their aims was to find out where people nearby obtain their information about this. Overall, this analysis shows the local people are not really engaging, and this engagement, up to when this paper was published, had been in decline. The most actively followed source of information, the nuclear operator TVO's newsletter at 18% is likely this high as it has been delivered to all in the area free of charge four times a year since the 1970s. Newspapers are also prominent with 94% either actively or occasionally using these to obtain information on the GDF, with a similar dynamics with television and radio. The next significant source is information from other people – friends, relatives, etc. Kojo et al. (2010, p. 173) claim this relationality based on "an image of the expertise of Posiva and its personnel, likewise their capabilities, honesty and the predictability of activity" is also used to assess the GDF. The other issue Kojo et al. (2010) and Kari et al. (2010) discovered in their surveys is what they call a community divided. This involves the familiar dynamics of wealth, power, and privilege. This involves men more trusting and supportive of the GDF over women, those with higher incomes and more highly educated more trusting and supportive of the GDF, and those supportive of the political right and centre more trusting and supportive of the GDF, with the Left Alliance more divided on this (Kari et al. 2010; Kojo et al. 2010). This tends to support the argument made earlier, a media that takes a pro-nuclear position and tends to take the industry's side is not doing its job if this job is to report on polemical issues to all the population independently (Sen 2009).

Intragenerational and Intergenerational Justice and Finland's GDF

This book concerns the ethics of intragenerational and intergenerational justice and argues these should be seen as complementary on a principle that justice in one place/time *at the expense* of another is a contradiction. A crucial part of this endeavour is asking people what they think and deliberative activity from the bottom-up for practical reasons – in terms of social acceptance and trust – and substantive reasons – this is what should happen in democratic societies (Bell 2021; Cotton 2017). This argument has been made in the Finnish nuclear waste context by Vilhunen et al. (2022), who using Blowers' (2010, 2020) peripheralisation framework examined the views of two Finnish communities on the GDF plan. The findings and conclusions of this study and the relevant parts of the previous surveys discussed earlier (Kari et al. 2010; Kojo et al. 2010) will be discussed in the final section of this chapter.

Vilhunen et al. (2022) compared two communities at different stages of Blowers' (2020) peripheralisation process – a nuclear oasis and a greenfield site. Briefly, an oasis is a nuclear community that has been long established with a NPP and/or waste stored for the interim for many years with the five aspects of peripheralisation quite distinct. A greenfield site is relatively new to nuclear infrastructure with less socialisation and aspects of peripheralisation established (Blowers 2010, 2020). The two communities surveyed are Eurajoki – the nuclear oasis – and Pyhajoki – the proposed site of the new NPP being constructed by Fennovoima. Vilhunen et al. (2022) compared the two communities in terms of the five aspects of peripheralisation. Geographically both places are fairly remote with low population density and at least 15 kilometres from any place with more than 25,000 people. Both places have aspects of economic marginality with high unemployment rates (12% and 12.6%, respectively); however, Eurajoki, as the more established nuclear community, is more dependent, with the local government dependent on taxes from the industry and the local community dependent on the jobs as TVO and Posiva are the biggest private employers in the area. Both places have the right of veto (The Nuclear Energy Act) within the municipality, so this can be seen as a form of political power. However, this power is firmly vested with the politicians in each municipality, and requests for extending this power via a referenda in both locales have been ignored. Also, beyond saying yes or no, the municipality has little say in other key aspects of the GDF, for instance regarding the inventory or what goes into it. In terms of cultures of acceptance, both places show high levels of trust in STUK, and 47% in Eurajoki and 57% in Pyhajoki believe nuclear energy is ethically justified. However, Pyhajoki does have an active anti-nuclear movement. Finally, there have been no nuclear events or historical dumping, so the existent risks are fairly low, but both communities have expressed concern on the effects their NPPs will have on sea life and fishing (Vilhunen et al. 2022).

Vilhunen et al.'s (2022) study of these two Finnish communities finds information to support that reported by Kari et al. (2010) on divided communities, but also adds crucial aspects of intragenerational and intergenerational justice. Firstly, in both communities men are more supportive than women on accepting the waste from the new Fennovoima NPP in their municipality; however, for both men and women, this is a minority position, with overall 68% in Pyhajoki and 64% in Eurajoki opposing this. This suggests at least some opposition to creating new nuclear waste. A further interesting finding was majorities of 50% in Pyhajoki and 57% in Eurajoki support a municipal referendum on the issue. An aspect of intragenerational injustice was evident, with 40% of both communities believing that there was not enough involvement of the society's disadvantaged in the process. Intergenerational aspects were somewhat in tension with overwhelming support (85% in both places) for dealing with and deciding on a GDF now, but with 40% in Pyhajoki and 38% in Eurajoki thinking disposal of nuclear waste in Finnish bedrock is not fair on future generations. Vilhunen et al. (2022, p. 1040) conclude that the nuclear industry "should acknowledge ethical diversities at local level and consider their evaluations and assessments more seriously from the viewpoint of ethics." They proceed to cite Cotton (2018) who argues for what he terms scalar parity. This involves a participatory and voluntary decision-making process that can "simultaneously scale up local community concerns, and scale down national concerns." This allows the rebalancing of the local, regional, and national and the amelioration or resolution of environmental injustices surrounding wicked problems (Cotton 2018, p. 253). A first stage for this in the Finnish context might be to allow for more participatory processes leading up to referenda in order to broaden and deepen the democratic process. This is arguably crucial in nuclear issues, even in countries with high social trust.

Conclusion

Again, there is an asymmetry between the rhetoric and the performance of the NWMOs, in this case in Finland. This asymmetry is perhaps less overt than the Canadian case, and this is arguably because the Finnish NWMOs make less expansive claims. For example, in STUK's four values, it claims maintaining up-to-date scientific knowledge in order to understand complexity, to be transparent and open to new ideas, to have the courage to try things differently, to accept and learn from mistakes, to work cooperatively, and to value and respect everyone's work. These values are not as progressive as the Canadian NWMO's social and ethical framework. However, STUK and Posiva still manage to undermine themselves both in their own assumptions and in the GDF implementation.

The major problematic assumption is the idea of the *correct* understanding of risk that STUK aims to communicate to the public. This is presumably the probabilistic risk assessment or formal risk assessment that STUK presents as one of its responsibilities. Apparently, this up-to-date scientific understanding does not include the voluminous literature on risk. Firstly, this literature specifically questions the validity of formal risk analysis in the way it simplifies causality, agency, and uncertainty (Jasanoff 1999). Posiva openly admits that it uses a "simplified dose calculation" (simplified causality), claims it has assessed its safety case against "pessimistic" scenarios (simplified agency of both people and ecosystems over 1 million years), and claims to have found and delivered a "safe" solution in the GDF (a simplification of inherent uncertainty over 1 million years). Thus, formal risk analysis is *a* tool for understanding risk, not the *correct* tool, and, more importantly, lay people tend not to use this or act on it. Instead, they employ various other cultural heuristics including consideration trust in institutions evaluated on both how they acted in the past and the individual/groups' worldview (Kasperson et al. 1988; Sturgis & Allum 2004). It also involves communicating with family, friends, and colleagues (Plough & Krimsky 1987); questioning the morality or legitimacy of the distribution of risk; being more precautionary than their "Promethean" policy-makers (Dryzek et al. 2008); and a whole host of socio-psychological amplification and attenuation devices (Kasperson et al. 1988; Alhakami & Slovic 1994; Plough & Krimsky 1987). To elide this complexity with a specious notion of *correct* understanding is clearly not being courageous, scientifically up-to-date, or even transparent.

In the search for and implementation of the GDF, there were a number of problematic elements that can be collectively described as a democratic deficit and peripheralisation. There is also a connected aspect of the energopower of the nuclear. Eurajoki's municipality's change of direction seems to be motivated by its economic reliance on TVO; thus, this is evidence of economic marginality and the monopsony power of the operator. This energopower is by-directional in that it takes social power to place and operate a NPP, but this NPP and its profits over its amortisation period reinforce this social power. Geographically, Eurajoki is clearly distant from the major urban centres of Finland, while the benefit packages negotiated privately and being comparatively meagre speak of political powerlessness – both the community excluded and the politicians who negotiated such poor terms. Finally, the one-sided pro-nuclear media coverage, in contrast to the gendered, wealth, and politically divided community in Eurajoki and Pyhajoki, implies democratic deficits. Overall, Finland will likely be the first to achieve a GDF solution, but this process and its guiding assumptions were far from perfect. The question is what lessons will Finland's NWMOs try to sell to the world. Will this be the lesson of first mover, this is the best that can be done with such a divisive sociotechnical assemblage, or this is where we went wrong and where the process might be improved?

8 The United Kingdom and Nuclear Power and Waste

This chapter will consider the case of the United Kingdom, which, like the previous cases, is a bundle of contradictions but all particular to this particular state. Like the previous chapters, we will use the core nuclear concepts introduced in Chapter five to assess the intra- and intergenerational aspects of nuclear power and waste in this country. The first section will cover the United Kingdom's nuclear history. The second section will examine the aims, objectives, and underlying assumptions of the Nuclear Waste Management (NWM) authorities. Thirdly, we will examine the actual practices of governance, the achievement of stated objectives, and the cost and efficiency of UK nuclear technopolitics and energopower. Finally, we will discuss the democratic credentials of nuclear technopolitics in the United Kingdom generally. The overall argument will be that while the nuclear waste processes are laudably deliberative, this is being undermined by an implicit support for the wider nuclear industry and the government's plans for this industry's significant expansion. We can frame this as a democratic deficit, which was present in both previous cases but will be expanded on in the final section of this chapter. This is the contradiction embedded in deliberative and democratic engagement at the waste end/end-of-pipe side, but little or no deliberative democratic engagement at the nuclear production or supply side. We frame this as the democratic deficit, so far inherent in the energopower of the nuclear.

Context and History of Nuclear Technologies in the United Kingdom

In the United Kingdom, the management of nuclear waste has a subtly different history to that of Finland and Canada due to two factors: firstly, the intimate connection between the civilian and military uses of nuclear technology since its inception in the early 1950s and, secondly, the domestic and geopolitical context of fossil fuel use in the United Kingdom during the period of 1950–1980. It is the combined sociotechnical imaginary of nuclear power, both as a public technology (Trischler & Bud 2018) serving the national interest and as a source of military resources serving national security objectives,

DOI: 10.4324/9781003422686-11

that led to the path dependency of the nuclear industry's expansion, decline, and potential rebirth in the United Kingdom.

From a waste management perspective, Thomas (2023) describes how the UK nuclear industry over the past 50 years can be characterised as treating the disposal of nuclear waste primarily is a social problem rather than a technological problem. Yet despite this, very little progress has been made towards the siting of a deep geological disposal facility for the management of high-level and intermediate-level nuclear wastes. Also, the question of what to do with the plutonium resulting from the re-processing of spent fuel remains subject to a final political decision. We present here the history of nuclear waste management in four emergent periods associated with concurrent discourses. These are somewhat circular, moving from securitisation as a military discourse, towards the opening up of technological decision-making through what is commonly called the "deliberative turn" emphasising participative justice, and then a turn back towards securitisation under conditions of climate emergency (Cotton 2020). These four historical periods and accompanying discourses are detailed in Table 8.1 and described in further detail in the next section.

Period 1: Economic and Military Securitisation

The first period of UK nuclear history involved the country's involvement in the development of the nuclear weapons programme during and after the Manhattan Project. This drive to develop nuclear weapons meant that military needs took primacy within nuclear research and development. As such, civilian and military nuclear technologies are inherently entwined. This has significant ramifications for radioactive waste management practices extending from the inception of a military-industrial sociotechnical pathway for nuclear power in the early 1950s up to the present day. There were three significant events that we highlight during this first period as well as detailing problems with nuclear technology. These technological problems often become sanitised or occluded by stories of technological progress (Verma 2021), as we will see later.

The first event was on 3 October 1952, when the UK government detonated its first test of a hydrogen bomb. This was launched from *HMS Plym* anchored 600 metres off the coast of Trimouille Island in the Montebello Island region, approximately 80 miles off the northwest coast of Australia (Spinardi 1997). Further tests were then conducted by the United Kingdom at Easter Island, Emu Field, and Maralinga (Smith 2016). The UK government's use of its colonial power in the Pacific Ocean to test nuclear weapons is itself a form of global environmental and social injustice (Ruff 2015), leading to significant humanitarian and environmental repercussions. It is clear from historical analysis that the UK government still has insufficiently addressed the needs and rights of test survivors and has proved wholly inadequate in

Table 8.1 UK Nuclear History Stages and Characteristic Discourses

Historical Stage and Dates/ Discourses	*Description*	*Blowers' Discourse Detail*
1. **Economic, energy, and military securitisation (1940–1976)**	Nuclear weapons were the original objective with energy a by-product and subsequent symbiotic relation between the two. In this period, nuclear was industry characterised by secrecy, occlusion of problems, and lack of democratic oversight	***Discourse of trust in technology*** The time of the "peaceful atom," promises of energy too abundant and cheap to meter. The 1950s were a time of austerity but also social provision by the welfare state and so trust in political institutions and technical innovation
2. **Nuclear energy expansion and the recognition of waste as an environmental concern (1976–1997)**	Nuclear waste is recognised in the context of environmental and radiological protection. The Flowers Report in 1976 argues that expansion of the nuclear industry could not continue without a waste solution. The government and industry over the next two decades attempt decide–announce–defend top-down solution which all fail	***Discourse of danger and distrust*** Expert fears of health and environment impacts emerge spurred by disasters in the United States and USSR. Anti-nuclear movement emerges combining with civil society to oppose the industry's expansion, implying a shift in the balance of power. With the Cold War's end and the memory of disasters fading, a more cooperative context emerges in which opposing camps come together to try to deal with the issue of nuclear waste
3. **The Deliberative Turn (1997–2007)**	New Labour government in 1997 sets up the Managing Radioactive Waste Safely (MRWS) programme, which sets up the Committee on Radioactive Waste Management (CoRWM). CoRWM conducts one of the most deliberative democratic exercises in UK history and establishes the solution a geological disposal facility (GDF) and that this should be voluntarily accepted by the host community	***Discourse of consensus and cooperation*** New Labour defined as a period of democratic renewal and experimentation in new forms of decision-making through direct dialogue with citizen actors

(*Continued*)

Table 8.1 (Continued)

Historical Stage and Dates/ Discourses	*Description*	*Blowers' Discourse Detail*
4. **Climate change securitisation (2007–ongoing)**	New Labour makes nuclear power part of its climate change policy. This move to securitisation undermines some of previous deliberative progress and more broadly established justice principles (Aarhus Convention), with a UK judicial review finding the government failed to consult the public properly on this decision, a process mirrored in the Netherlands Subsequent Coalition Government and Conservative Government action to rebuild nuclear raises questions around infrastructure cost (including the strike price of nuclear power), and national security through the involvement of Chinese industrial investment and technology transfer to the UK nuclear sector	***Discourse of security*** A discourse that uses the metaphors of conflict, e.g. the "war on terror"; scarcity, e.g. "peak oil"; and urgency, e.g. ecological modernisation in an era of climate emergency. Terrorism and climate change compete for governance attention but combine in the notion of energy security. Nuclear power here is framed as the solution to the energy trilemma of energy security, affordability, and action towards net zero

Source: Adapted from Thomas (2023); Cotton (2020; Blowers (2010)

tackling the ongoing environmental concerns in the marine and terrestrial environment (Prăvălie 2014; Alexis-Martin et al. 2021). This fundamental intra- and intergenerational injustice concerns the distribution of radioactive hazards and risk. The most vulnerable populations are women, children, and the socially marginalised, particularly women who were pregnant or about to become pregnant during this testing period. Each of these groups manifest disproportionate negative health impacts from exposure to radioactive contaminants (Heath 2013).

The second event occurred in 1957 with Egypt's nationalisation of the Suez Canal and the resultant military intervention by the United Kingdom, France, and Israel. This exacerbated energy security concerns for Western Europe following the disruption to oil supply. Under these political conditions, the promise of "hard power" gained from becoming a nuclear weaponised state (see Baylis & Stoddart 2012) offers to bolster the geopolitical standing of a war-shattered British economy. This combines with a growing international regime of nuclear governance (Atoms for Peace in 1953) and concerns over energy supply, creating a sense of political urgency towards nuclear expansion in the United Kingdom. This urgency led to a policy discourse of securitisation in the face of geopolitical threats (the rise of the Soviet powers after the Second World War), economic threats (including commercial competition from US reactor designs), and domestic energy security threats (the National Union of Mineworkers – see Mitchell 2013). We see, therefore, a complex combination of security, economic, and geopolitical factors that prompted the government to adopt a policy strategy of nuclear expansionism and technological boosterism – a combination of political support, intensive research and development, and the overlapping interests of military and commercial actors in defining the energy and industrial strategy of the 1950s and 1960s. There was therefore a curious combination of open political support for nuclear reactors to be built, while all the operational aspects took place under conditions of administrative centralisation and high levels of secrecy (Rosa & Freudenburg 1993). The programme was led by a special department within what was then the Ministry of Supply, given the responsibility for civilian research and development, and the production of both fissile material and, ultimately, a functioning hydrogen bomb (Gowing 1974). We see therefore a mix of boosterism, hype, and covert action taking place simultaneously in Cumbria in Northwest England, exemplifying the complexity of nuclear politics from its inception in post-war Britain.

The third event regards a fire at the first civilian reactor in 1957. The first civilian reactor in the United Kingdom called Windscale entered service in 1956. The construction of the Windscale Piles (reactors) for military plutonium production in Cumbria in Northwest England was the first major investment in military–civilian nuclear technology production. However, it was also the site of the first major nuclear industry incident. The Windscale Piles incident, a major nuclear accident, happened at the Windscale Works site in Cumberland (now Sellafield in West Cumbria), in Northwest England, on 10 October 1957. It was a crucial event in the history of nuclear safety; the environmental impacts of this event were significant and long-lasting, leading to advancements in safety protocols and the understanding of potential risks associated with nuclear reactors. On 10 October 1957, the operators were carrying out maintenance on Pile 1, which had been in operation for about a year. During routine maintenance, an issue arose with the air-and-water cooling system. To address the cooling problem, the decision was made to perform a "Wigner release" – raising and then lowering the power level of the reactor, to try and get the distorted graphite blocks back into their proper shape. This was expected to help maintain cooling. However, as the power increased, an uncontrollable rise in temperature occurred in the reactor core. The physicist reading the instruments found that the effect was not complete and would, he thought (incorrectly), require more heat. He then withdrew additional boron control rods, which led to the ignition of the graphite moderator surrounding the uranium fuel (Dalquist 2004). The resulting fire spread rapidly through the graphite core, releasing radioactive iodine and polonium into the environment. Efforts were made to extinguish the fire using water and firefighting foam. The decision to use water, however, contributed to the release of radioactive steam into the atmosphere. Once the scale of the incident became apparent, an alert was issued, and the surrounding area was evacuated due to the risk of a potential explosion. The Windscale Piles incident released a significant amount of radioactive material into the atmosphere, leading to contamination of the surrounding area. Interestingly, the Windscale piles were fitted with expensive filtering equipment. At the time they were referred to as Cockcroft's Folly, as the filtering was not deemed necessary at the time of their construction. However, with the Windscale fire raging for three days, the presence of the filters prevented as much as 95% of the radioactive material produced from entering the atmosphere. Therefore, this foresight averted a much broader environmental disaster (Arnold 1992). Nevertheless, the government prohibited the sale of milk from a 500-square-kilometre zone around the reactor site for several weeks, diluting it a thousandfold and dumping it in the Irish Sea. The overall health and environmental impact of the incident was then monitored over subsequent years, with growing concerns mounting in the local community over the overall safety profile of nuclear power in the Cumbrian region.

Thus, the first period of techno-optimism began over time to be undermined by concerns over health, public safety, and environmental impacts of nuclear technology. However, policy-makers' desire for the expansion of nuclear power were undeterred, and governments engaged in the now familiar projects of information control and suppression. This included providing only very limited information about the Windscale fire while downplaying its severity. The reactor at Windscale that had the incident was sealed and not opened until a clean-up operation began in the late 1980s. Thus, this information control, combined with aspects of peripheralisation, or the geographical remoteness of the nuclear communities and their associated political powerlessness, allowed authorities to frame the Windscale fire as a local and limited event. This is in stark contrast to the high-profile nuclear incidents such as the Chernobyl, Three Mile Island, and Fukushima disasters, which made international headlines. We also see in this period aspects of nuclear colonialism in the use of colonised land and livelihoods as nuclear weapons testing grounds. However, these events were also suppressed and/or difficult to uncover. Therefore, with little in the way of public opposition, arguably because of the success of the information control, the nuclear programme continues unabated into the next decade.

Period 2: Nuclear Energy Expansion and the Recognition of Waste as an Environmental Concern

The 1960s saw nuclear development in the United Kingdom and wider world. Scientific and technical achievements (and indeed nuclear scientists themselves) were looked upon with favour by the media and became highly regarded within energy policy networks and by the UK public at large. By 1960, there were 17 nuclear power reactors in operation, with a total electrical capacity of 1,200 megawatts (MWe) in four countries: the United Kingdom, France, the USSR, and the United States. Nuclear power programmes had been launched in another six countries, electric power utilities were placing their orders for nuclear plants on a routine basis, and by 1970 there were already 90 nuclear units operating in 15 countries with a total capacity of 16,500 MWe (Char & Csik 1987). However, underneath this sunny picture were fundamental problems with the United Kingdom's reactor technology and a growing nuclear waste problem.

By the 1960s, nuclear technology was expanding across the world, driven by Westinghouse reactor designs from the United States. However, as Thomas (2023) notes, in the United Kingdom, the picture was more complicated. Firstly, the Magnox design was abandoned as no-longer commercially competitive, to be replaced with a tweaked version: the Advanced Gas-cooled Reactor (AGR) design. However, this too was faced with cost over-runs and delays, culminating in the construction of five expensive and unreliable reactors. Thomas (2023, p. 203) describes these designs as the most "delayed and unreliable set of reactors built in the world." Subsequent attempts at government reactor choices, including the Steam-Generating Heavy Water Reactor, dual AGR/Pressurised Water Reactor (PWR) policy, and the PWR programme, also broadly failed. Only two more AGRs and one PWR (Sizewell B) were completed in 1989 and 1995. The fast breeder reactor programme, first driven by concerns over uranium availability and then driven by a sociotechnical pathway of plutonium production from reprocessing, remained influential in nuclear new build policy, but it too was later abandoned by the 1980s (Thomas 2023).

During this tumultuous period of different reactor designs and nuclear securitisation drivers, the long-term problem of waste was largely downplayed or ignored by nuclear authorities (Welsh 1993). Although part of the newly created UK Atomic Energy Authority (UKAEA)'s remit was to investigate waste management strategy as Hookway (1984, p. 123) states,

> the early stages of our nuclear programme had hanging over them an urgency which led to the shelving of problems if this were possible. At Windscale . . . the highly active liquors [HLW] were kept in tanks without real thought being given to their eventual fate. Intermediate level wastes were put into silos, little more than crude concrete boxes without there being any plan to retrieve them, let alone process and eventually dispose of them.

Between 1952 and 1976, there was minimal consideration regarding the disposal of highly active wastes generated by military and nuclear electricity programmes. The policy in this field lacked coherence and consistency, with successive governments relying on different interpretations of optimal scientific advice and being influenced by the perceived necessity to safeguard spent fuel reprocessing at all costs (Berkhout 1991; Cotton 2017). During the early 1950s, an extensive excavation was undertaken along the Cumbrian coast, and a trench was subsequently encased in concrete. Dubbed B30, and equivalent in length to three Olympic swimming pools, this structure was constructed to accommodate large quantities of spent nuclear fuel. The Magnox programme's spent fuel assemblies were intensely radioactive. The rods were simply deposited in B30 and submerged to allow for cooling. These were originally only supposed to be left in B30 for a period of three months before undergoing reprocessing. However, during the miners' strikes in the 1970s and the Winter of Discontent, coal power stations were shut down, leading to an increased dependence on nuclear plants to meet the drop in electricity supply. This shift resulted in the generation of spent fuel to much faster rate than could be reprocessed. The silos and ponds, originally constructed to prevent airborne contamination in case the fuel dried out, quickly reached their capacity. Because fuel rods were not being taken out fast enough, they corroded in the water and broke down into a radioactive sludge. The sludge and the contamination of B30 and other sites across Sellafield remain an expensive and dangerous environmental remediation operation today (Gray et al. 1995; Lawson & Isaac 2023).

Reprocessing involved chemically and mechanically separating uranium, plutonium, and heat-generating fissile products from the spent fuel assembly. The UKAEA had begun experimental work on waste vitrification – storing the waste in a glass-form waste package that was chemically non-reactive and waterproof. The waste products remained heat producing, so the policy was to store the HLW on-site in temporary facilities to allow for heat dissipation. This was seen at the time as a short-term solution but allowed nuclear authorities to continue producing nuclear power and reprocessing spent fuel without making firm plans on a waste management solution. As such, they remained largely silent on the long-term technical and socio-economic impacts, and health and environmental risks of the waste management process. Similarly, decisions on the heat-generating products of spent fuel reprocessing remained unresolved during the 1960s and early 1970s. These processes and their guiding decisions result in a number of intra- and intergenerational injustices. The public were unaware of the scale, cost, and environmental risk associated with the re-processing facility and spent fuel storage, nor were the public privy to the wholly inadequate disposal mechanisms for intermediate-level wastes (ILWs) during this time. Secrecy and the political suppression of dissenting voices fundamentally reduced the autonomy and decision-making capacity of the UK public to make informed choices about nuclear futures during this period. The securitisation framing of the nuclear technology project meant that the sense of urgency for nuclear expansion combined with the political protection of reprocessing activity became state secrets. Justice in relation to nuclear technologies was defined in utilitarian terms as a matter of national interest and of public good. Alternative ethical considerations, such as ecological justice, fairness for future generations, and procedural justice in nuclear decision-making, were simply not part of the nuclear political discourse in the 1950s and 1960s. Civil society actors were deeply limited in their opportunity to put forth alternative ethical positions, in part because they were unaware of the scale of the problem before them.

Despite these problems, for spent fuel, HLW, and plutonium, there was at least something resembling a plan to begin storing these materials safely after reprocessing. For other wastes, including radioactively contaminated cements and other LLW and ILW, these were just dumped. ILW cannisters were deposited at different sites in the Atlantic Ocean starting in 1949 and in the Irish Sea in the 1950s. The underlying principle was to dilute and disperse radioactivity within such a large body of water, thus spreading the overall risk to the marine environment so that it would be indistinguishable from background levels of radiation. LLWs were also dumped in the English Channel between 1950 and 1963. Between 1949 and 1982, an estimated 33,000 cubic metres of radioactive wastes were released into the Atlantic and British coastal waters (Cotton 2017). Data submitted to the International Atomic Energy Authority indicate that the United Kingdom engaged in 34 operations across 15 sites in the Atlantic Ocean, dumping 74,052 tonnes of nuclear wastes. Although the mantra amongst nuclear authorities was to dilute and disperse, this combined action of ocean dumping led to an overall additional radioactivity load of 3.51×10^7 gigabecquerels into the marine environment (IAEA 1999). It is notable perhaps that UKAEA stated at the time that sea disposal was not a viable long-term solution to the waste problem, but they were nevertheless slow in defining and implementing an alternative solution for ILW and LLW (Saddington & Templeton 1958; Van Dyke 1988). Thus, unnecessary environmental contamination, and hence greater expense borne in the future to clean up such contamination, was a core feature of this period of nuclear activity. The abject failure to plan ahead raised significant intra- and intergenerational injustices concerning the irradiation of the Cumbrian coastline, leaks of radioactive materials into the biosphere in the west Cumbrian environment, expensive clean-up operations across the Sellafield site, and risks to nuclear workers from unnecessary exposure.

By the mid-1970s, the United Kingdom saw bigger changes in the environmental discourse that prompted greater action towards nuclear safety and a long-term management solution for nuclear wastes. It is notable that, during this period, the civilian and military aspects of nuclear at least became nominally separated. Formed in the 1970s, the establishment of British Nuclear Fuels (BNFL) bolstered the role of private industry in the field of nuclear development. During this period, the UKAEA directed its attention towards the production of civilian power, while the Ministry of Defence assumed responsibility for weaponry.

In 1970, the Royal Commission on Environmental Pollution (RCEP) was set up to look into issues of environmental pollution and impacts on socio-environmental health. The RCEP produced a report in 1976, also known as the Flowers Report, which recommended no new nuclear programme should commence until it was shown "beyond reasonable doubt" that there was a solution to nuclear waste (RCEP 1976). This arguably forced the nuclear industry to start taking the waste issue more seriously. Initially, the Department for the Environment was tasked with the responsibility of finding a permanent disposal site. Extensive geological research then began to identify suitable sites for a deep geological repository for HLW. In 1979, the first test drilling took place at Altnabreac (in Scotland) and Harwell (in Oxfordshire) (Mather et al. 1979), and public inquiries were conducted regarding similar drilling proposals in the Galloway Hills, Cheviot Hills, and Leicestershire as potential locations for HLW disposal (Nirex 2005). Yet drilling was halted in 1981 in each case primarily due to strong public opposition to proposals in light of the decide–announce–defend (DAD) nature of the decision-making process. DAD involves government-led radioactive waste management organisation making a decision on the site, communicating this decision, and then attempting to persuade a hostile community to accept this. However, this often provokes place-protective action in the form of protest, political campaigning, and local council opposition and obstruction.

The 1976 Flowers Report had called for an independent radioactive waste management authority, independent of government and industry interests. However, this was not the approach taken. In 1982, the Nuclear Industry Radioactive Waste Executive (Nirex) was established as a private company by the bodies within the British nuclear industry, including BNFL. The fact that Nirex was a commercial rather than a public entity was significant as this diminished public trust in the process of site selection. This was because it gave the impression that Nirex as a company was tasked with dealing with the tail-end pollutants of its industrial shareholders. Nevertheless, Nirex was tasked by the government with finding

- a deep disposal site for long-lived ILW in an underground repository and
- a shallow burial site for LLW and short-lived ILW in engineered trenches 20–30 metres below the surface.

What follows was a series of failed attempts by Nirex to achieve this task partly due to its top-down technocratic approach. This culminated in the 1990s with Nirex's attempt to establish a Rock Characterisation Facility (RFC) in Cumbria. This begins in 1989 when the Secretary of State for the Environment announced that Nirex's plans for a "deep repository" must be on land. Subsequently, Nirex announced it would focus its search on two nuclear communities – Sellafield in Cumbria and Dounreay in Scotland. The government responded saying further work needed to be done before a specific site was selected; however, Nirex went ahead with the investigation in Cumbria in 1991. Critics of the process, including the geophysicist David Smyth, argued that the process of selecting Longlands Farm near Sellafield was manipulated from the start – the Longlands Farm selected site was indeed near Sellafield but was very different, geologically speaking, from the "Sellafield" listed in the original list of 437 potential sites. Smyth (2012) later argued that the criteria were manipulated and the site location moved twice to ensure that a near-Sellafield location was chosen for political rather than technical reasons. As well as drilling boreholes, Nirex proposed to excavate a rock characterisation facility called the "Rock Lab" or RCF. This would be at a depth of more than 650 metres, consist of a set of shafts and galleries, and be used to help develop the safety case for the repository through testing of rock strength and hydrogeology, in advance of the full repository proposal. Nirex claimed RCF would be purely an experimental facility. However, the Cumbria County Council and environmental non-governmental organisations including Friends of the Earth and Greenpeace were sceptical. These critics argued that the RCF was integral to Nirex's repository plans and thus would be the first stage of a GDF.

On 1 February 1994, Nirex submitted a planning application for the RCF to the Cumbria County Council. The council refused this planning application on 20 December. It must be noted that this was separate from a planning application for the repository itself, for which the government had promised a full public inquiry. By 27 February 1995, the then Environment Secretary John Gummer announced that a planning appeal by Nirex for the RCF could be considered by September 1995. This took place and the planning appeal closed on 1 February 1996. There was however uncertainty at the time about when this decision was going to be released, and questions in the parliament were given vague answers. When it came to the appeal itself, the inspector overseeing the Public Planning Inquiry of 1995–1996 for Longlands Farm invalidated Nirex's position, contending that the RCF represented a preliminary phase towards establishing a deep geological disposal facility. He also criticised the site selection process, deeming the geological and hydrogeological characteristics of the site as unsuitable and advocating for the consideration of other more geologically suitable sites across England. Thus after an estimated £400 million in costs, one of the last acts of the outgoing Conservative government was to reject the planning application for the site (Haszeldine & Smythe 1997).

Period 3: The Deliberative Turn

The incoming New Labour government took a very different approach to its predecessor. There was a clear shift towards greater consideration for the social, political, ethical, and psychological aspects of radioactive waste management policy-making. This shift was in light of Nirex's planning failures, alongside broader changes to the ways that citizens and government institutions interacted with one another. In the early 2000s, there was concerted effort to involve UK publics in various stages of policy development on a range of different issues. This period in the early 2000s is sometimes described as a deliberative turn in policy-making practice – a pattern seen more broadly across the European Union with an emphasis upon Responsible Research and Innovation through the Framework Programme of scientific and research and development funding, and a growing interest in new forms of democratic engagement through citizen's juries, consensus conferences, and E-government platforms. However, the focus on deliberative democratic engagement was particularly notable in the governance of what Cotton (2014) termed socially and ethically contentious technologies which included genetically modified organisms (notable in the GMNation! Public engagement programme), and later radioactive waste management. Moreover, the pursuit of the Millennium Development Goals and later the Sustainable Development Goals saw significant changes, notably with the institutionalisation of public and stakeholder engagement across government departments (Chilvers & Burgess 2008). This broader commitment to improving the deliberative capacity of heterogenous publics to be involved in radioactive waste also involved significant learning from Finnish, Canadian, and Swedish cases in the design of community engagement mechanisms, voluntarism, and stakeholder communication practices (Cotton 2017).

More specifically, this involved the establishment of a new Committee on Radioactive Waste Management (CoRWM) in 2004. The CoRWM was tasked with overseeing a thorough examination of radioactive waste management options. The overall aims of the committee were to

- identify relevant areas of scientific knowledge and uncertainties;
- to learn from international experience;
- to consider ethical issues including intergenerational justice;
- and to design and implement an engagement process with the public and with relevant industry, scientific, and community representative stakeholders.

This process began with a long list of potential radioactive waste management options including more esoteric options such as burying waste in ice sheets or subduction zones or blasting it into space. Their evaluation process to whittle down this long list of options integrated various factors, including multi-attribute decision analysis, independent ethical assessment, and extensive engagement with the public and stakeholders (CoRWM 2006). This engagement involved a range of activities such as a national stakeholder forum, selected discussion groups, citizens' panels, a schools project, and various public meetings across nuclear sites. The approach emphasised accessibility to scientific information while ensuring it did not dominate decision-making. The operating phrase was the George William Russell phrase of "science on tap, not on top" and was in stark contrast to the previous technocratic approach employed by Nirex (Cotton 2017).

This wider deliberative engagement programme had many positive aspects involving public participation, education, and debate (Cotton 2017, 2020). However, one critique of the overall programme was the relative weight and influence of expert versus the public, with some arguing for the whole process favouring certain types of technical expertise (Chilvers & Burgess 2008). This is problematic for, as argued earlier, the what/how technical questions should be preceded by philosophical, moral, and ethical questions of why and should we (O'Riordan 2000; Habermas 1972). Ethics is morality considered and reflected upon, something that deliberative democracy enables for all (Dryzek 2000, 2013).

The CoRWM's final report made a number of important recommendations. It chose as the best option a GDF but stipulated that this must follow a voluntary process whereby a host community chooses to come forward to learn about the issues but has the right to withdraw. As will be discussed later, these aspects remain as important features of democratic safeguards against top-down technocratic governance. However, before the Conservative and Liberal coalition and then Conservative administrations began their work undoing elements of this deliberative turn, a crucial problem had already emerged in UK nuclear governance.

Period 4: Climate Change Securitisation

The fourth phase of climate change securitisation is arguably an example of how the term power should be used, as Malm (2016a) argues, in the double energy and social sense. The CoRWM's final conclusions on nuclear waste corresponded with the Flowers Report, in that they said:

> The main concern in the present context is that the proposals might be seized upon as providing a green light for new build. That is far from the case. New build wastes would extend the time-scales for implementation, possibly for very long but essentially unknowable future periods. Further, the political and ethical issues raised by the creation of more wastes are quite different from those relating to committed – and therefore, unavoidable – wastes. Should a new build programme be introduced, in CoRWM's view it would require a quite separate process to test and validate proposals for the management of the wastes arising.
>
> (CoRWM 2006, p. 13)

The Labour government was initially wary of the financial and social costs of the nuclear, but this position changed over time. A 2003 White Paper stated that nuclear power's current economics was "an unattractive option for new, carbon-free generating capacity," and as the waste issue was unresolved, the nuclear's use in the future should be subject to "the fullest public consultation" (DTI 2003, p. 12). A few years later, the rising price of oil and gas can be seen as a partial justification for the government's calls for a "nuclear renaissance" in 2006 and this becoming policy from 2008 (HM Government 2008). However, these rises in oil and gas prices do not justify or explain why nuclear was chosen over other options or why there was a lack of consultation on new NPPs. Greenpeace sued the government over this lack of deliberation, and the court found the government had not followed its own consultation processes and ordered it to go through this process again (Cotton 2017; Geels 2014). However, PM Blair vowed to ignore this and implement his nuclear policy regardless of this consultation outcome. Geels (2014) claims this was a result of the instrumental power of the nuclear lobby with direct access to the PM. In this reading, nuclear power has an energy and socio-political meaning.

There may be a more concrete explanation of the long-term governmental attachment to the nuclear industry despite its questionable economics. These questionable economics were recapitulated in the *Titans of Nuclear* (TN) podcast interview with financier David Sterns (TN 2022). In this interview, Sterns laments the finances of nuclear power, arguing that the excessive conditions of safety and regulation on NPPs were too costly and that the consumer should take on more of this risk and cost. Sterns also referred to big NPPs as a technology without a business plan and small modular reactors (SMRs) as a business plan without a technology (TN 2022). This does not sound like the finances of NPPs have improved so much in this "renaissance," a dynamic supported in relevant literature (Haywood et al. 2023; Price et al. 2023; Thomas 2023). This suggests that other factors may be at play with one answer offered by Stirling and Johnstone (2018; Johnstone & Stirling 2021). They argue that this strange attachment to nuclear by UK governments can be explained by the interdependency of the civilian and military aspects of the nuclear and can be seen through two aspects. Firstly, the knowledge and expertise between the civilian and the military are largely interchangeable, and states that want to keep the nuclear weapons need to maintain this social knowledge. This is openly talked about in other countries and in the United Kingdom's defence literature. Secondly, the technology itself is dual purpose, with all reactors designed to offload fissile material like plutonium-239, and supply chains, especially of uranium enrichment, are also shared, while specialist material for nuclear weapons like tritium is a by-product of NPP operations (Stirling & Johnstone 2018; Johnstone & Stirling 2021). This reading suggests the securitisation of the discourse and policy approach to nuclear power and climate change is securitisation in a very real and military sense.

Regardless of this reading, a basic problem remains with the United Kingdom's nuclear policy. This is that the Labour government and every government since have been using CoRWM's mandate for the GDF, while ignoring one of the crucial conditions. This was that the GDF was for unavoidable legacy waste, not avoidable new nuclear waste. This is a core tension that will run through this chapter and likely nuclear politics for the foreseeable future.

Current UK Nuclear Waste Policy

The main policy framework of nuclear waste management in the United Kingdom was developed from the CoRWM processes, set out in a White Paper in 2008 and updated in 2018 (BEIS 2018). In this updated policy framework, the government not only sets out its broad nuclear waste policy but also shows its underlying assumptions and thinking around the nuclear to which we shall return below. The main features of this policy are as follows:

- Classification of different types of waste and a description of the inventory to go into the GDF;
- Regulation of socio-environmental safety by the Office for Nuclear Regulation (ONR) and Environmental Agency;
- A "Working with Communities" consent-based process that involves "Government and its agencies working in partnership with communities" (BEIS 2018, p. 5).

This latter point can be elaborated upon. Any community or individual interested in proposing their area for the GDF can aim to set up a working group by contacting the Nuclear Waste Services (NWS), the government agency delivering the project. If both sides see possibility of progress, the principal local authority (district, county, or unitary authority) must be informed and the wider community involved in discussions. Then a working group can be formed, which must include a representative of the NWS, the original proposing group, and an independent chair and facilitator. The aim of this working group is to contact members and organisations of the wider community to set up a community partnership. This community partnership must include community members and organisations, the NWS, and an elected local authority actor. The aim of the community partnership is to provide information and answer the community's questions about the GDF process and community benefits of this project. This partnership is supposed to be diverse and represent and respect a range of opinions. The partnership is supposed to create a Community Partnership Agreement that is signed by all members of the group and set out the principles of working together, how decisions will be made, how roles and responsibilities will be shared, and how they will work with the process and engage the community. In terms of benefits, the government commits to providing £1 million annually for each community partnership, and this rises to £2.5 million when deep borehole investigation begins to assess suitability of the site. This all builds up to a Test of Public Support in order to decide if the community is willing to host the GDF. The community has the right to withdraw at any time before this test. The relevant local authority is to decide when the test takes place, and the community partnership decides what form this test takes (BEIS 2018). Overall, this is a good framework for a deliberative process. However, as ever, the problems reside not only in the unspoken and spoken assumptions but also in the actual implementation.

The NWS describes itself as specialists in dealing with nuclear waste, with a main goal of managing this waste to protect people and the environment now and in the future (NWS n.d. a; Gov.UK 2024a). This organisation commits to being a "safe and responsible business" and to be "customer and community focused" and aims to "develop capabilities." It further claims this community and customer focus of its business will place safety at its core, that it will be a great place to work where people will be respected, and that it will "deliver value for the taxpayer." The NWS claims it has an integrated approach that will ensure nuclear waste goes to the right place, at the right time, and at the right cost (NWS n.d. a). The tone of this organisation is very corporate, with buzzwords like *efficiency, innovation*, and *customer* throughout. However, it is the repetition of cost/s (repeated 13 times in NWS n.d. a) and value that seems most prominent and will be questioned later.

Finally, there are three familiar spoken and/or implied assumptions of this policy framework and associated agency. These assumptions are similar to those in previous chapters, which is not surprising as the NWS openly states it is in communication with NWMO in Finland and Canada, and like in Finland and Canada, its employees are drawn from a privatised energy business world. Firstly, the government minister states clear support for nuclear power, saying we as a country have benefitted from it for more than 60 years and it provides "clean energy to our homes and businesses and will continue to play an important role as we transition to a low-carbon economy" (BEIS 2018, p. 3). This neatly conflates the nuclear waste solutions with ongoing nuclear power as well as presenting both as almost inevitable public goods. Also, "clean energy" is a curious way to describe an energy technology that creates radioactive waste that will remain dangerous for over 200,000 years. The NWS is slightly more circumspect. It claims that nuclear technology has been "part of our lives" for over 60 years, providing just under 1/5 of our energy, and has been used for electricity, industry, medicine, and defence, but has also produced a legacy of waste which must be managed safely (NWS 2022). In another document, the NWS says "nuclear technology has delivered great benefits; supporting national defence and generating electricity for over 60 years" (NWS n.d. b, p. 5). Thus, while not as explicit in its support of nuclear power as the government, the NWS does implicitly support this industry. At the same time, the NWS here is engaging in Verma's (2021) linear narrative of technological progress, with these "great benefits" occluding the problematic history reported earlier. A second shared assumption concerns the safety case of the GDF. The NWS (2022) claims scientists around the world agree the GDF "will keep [nuclear waste] safe and secure over the hundreds of thousands of years it will take for the radioactivity to naturally decay." The problem with this statement is, firstly, that these scientists who say this cannot *know* this, as argued in previous chapters. Secondly, not *all* scientists agree and rarely if ever do, especially over things as highly uncertain as speculating over socio-environmental conditions over hundreds of thousands of years.

A final shared assumption between the government (BEIS 2018) and the NWS (NWS 2022, n.d.) is a very utilitarian belief that we are all responsible for this nuclear waste and will benefit from a GDF. This has elements that are true but becomes problematic when we dig down into the details. Are the people who have opposed nuclear power and weapons for the last 60 years also responsible? One could argue they probably used electricity from this source, but they would hardly have had any agency in choosing this, so it seems quite specious to argue this. Are the people who lived next to NPPs and did not benefit via employment, yet were likely to have faced higher risks, also responsible? And finally, the most salient issue surrounds what particular waste we as a society are responsible for. As the CoRWM argued, the legacy waste many would agree needs to be safely dealt with. However, new waste from a significantly expanded nuclear sector with new untested technologies is more polemical, and this is where the problems emerge. However, before we consider the consequences of an expanded nuclear power sector and waste inventory, we will firstly summarise the most recent developments in the search for the United Kingdom's GDF.

Implementation of the GDF

The search for the GDF is ongoing, and currently, there are three active community partnerships as well as some opposition to one of these partnerships. There are two community partnerships in Copeland in Cumbria – the Mid Copeland and South Copeland GDF community partnerships, and a third in Theddlethorpe in Lincolnshire. We will focus on this third place to explore the processes of the search for the GDF. These processes will be analysed using the meeting minutes from this Theddlethorpe Community Partnership and from an analysis of the local opposition to the GDF, the Guardians of the East Coast (GOTEC), and their website. The simple fact that these minutes are publicly available is a commendable move towards transparency that contrasts with the historical nuclear governance detailed earlier.

In the first meeting of the Working Group (involving a number of local councillors, NWS representatives, and local stakeholders), the parties established a number of objectives. These include recruiting more groups and individuals, as this is required to form the community partnership, to identify a suitable GDF site, and to develop a communication strategy (Theddlethorpe GDF Community Partnership, Working Group Meeting 1 2024a).[1] They point out that only 10% of people had engaged in the area, and they needed to involve the other 90%. For the Working Group to form a community partnership, it had to reflect the "different views of the community," which would entail involving the 90% of the "disinterested and/or unengaged" (WG 1). They then discuss how this should be a proactive process of going out into the community and should involve an age-targeted communications strategy with graphics, models, and an overall visual communicative approach (WG 1).

In the third and fourth meetings of the Theddlethorpe Working Group, the participants discussed working with "objector" groups. One of these groups, GOTEC is specifically mentioned, and it is decided a member of the working group will meet with this organisation (WGs 3 and 4). This meeting likely happened, and by the time the community partnership was established and had its second meeting in January 2023, a leading member of GOTEC and a leading member of GOTEC is present as a member of the community partnership. In the fourth meeting of the community partnership, a decision was taken to allow GOTEC and other local organisations to write short pieces in the local newspaper the partnership was producing and to have space at events (CP 4). Thus, we can contrast this inclusion of critical voices to the Canadian case in South Bruce in Chapter six, which seems to be less inclusive of critique and less diverse in who was included.

By the tenth meeting of this community partnership, in October 2023, it had produced a mission statement and a broad programme of action. The mission statement claims:

> To champion the interests of our community, local people, the environment and economy by independently encouraging the fullest participation in any decision to site a GDF in the Search Area throughout the process.
> (CP 10)

Other actions committed to are independence and impartiality, being open and accessible, and being locally informed (CP 10). In terms of independence and impartiality, the partnership often considers this problem and is keen to differentiate itself from the NWS and not be seen to be a "vessel" of this institution. In order to do this, the partnership aims to be transparent and open and refers to enabling online access to meetings, which does not appear to have begun yet. However, the community partnership's meetings are now open to the public, and there is an online forum to submit questions to these meetings (Theddlethorpe GDF Community Partnership 2024b). As to being locally informed, the partnership with the NWS appears to have conducted public surveys and done demographic studies in order to determine local needs (CP 3; CP 6). They are also conducting a visioning exercise asking the broad question "what does 'good' look like for our area?" (CP 2). Overall, the Theddlethorpe GDF Community Partnership seems to be meeting its statutory guidance to be a community-based, diverse body that operates in deliberative ways. However, there are questions about this specific process in Lincolnshire and about the Working with Community framework more generally.

There are questions about when the Test of Public Support will be held, and we could add questions of how this is conducted and who will be included and excluded. GOTEC (2024a) points out that Councillor Martin Hill said there had not been a referendum yet because people were not fully informed enough to make a decision. GOTEC compares this to the Brexit vote and how arguing then that people were not fully informed would not have been accepted. They also contrast the Theddlethorpe case to East Riding of Yorkshire Council, which voted to withdraw from the process, with even less information than available in East Lindsey, Lincolnshire (GOTEC 2024b). Thus, there are a number of issues here regarding an ambiguity that stems from the White Paper setting out the Working with Communities framework (BEIS 2018). This White Paper says the community can withdraw *at any time* until a Test of Public Support, but it also says that this decision of withdrawal will be taken by the principal local authority, as in the East Riding example. The principal local authority will also have the final say on when this Test of Public Support is taken (BEIS 2018, p. 6). Thus, the local authority seems to be the proxy for the community here, which creates a rather haphazard situation whereby however the local council feels about the nuclear, it can support and/or delay the Test of Public Support, or quickly veto the GDF process. Thomas (2023) frames this as easier to join the process than leave it, but with both the NWS and local authorities having veto power, it seems more arbitrary than this.

The method of this Test of Public Support is also vague, with GOTEC assuming it will be a referendum. If this was the choice, then there are questions of age and geographical inclusion and exclusion, which would apply to any other method used to test public support. Firstly, there are questions about the age requirements for voting in this Test of Public Support. As argued throughout this book, people should have a say on things that will directly affect them. The CEC and the GDF processes and events are going to affect today's children and young people much more than the older people currently making the decisions on these issues. Additionally, and practically, these children and young people will manage and/or oppose this intergenerational project in the future. Thus, this Test of Public Support is an opportunity to begin mitigate these generational injustices. There are also questions about the boundaries of any vote, with this also connecting to the community investment funding and infrastructure spending. This is the intragenerational aspect of this, with the previous rejection of a GDF in Cumbria based on the wider community concerns overriding the local nuclear community's desires. In contrast, GOTEC is concerned that the wider community may support the GDF based on better transport infrastructure but avoiding the local disruption and any problems emerging from the GDF (GOTEC 2024c). Thus, these tensions pull both ways and need to be carefully, fairly, and transparently managed. The next section turns to the inventory of the GDF, its costs, and the broader costs of the nuclear, which collectively could be far more transparent.

Expanding Costs and Expanding Inventories

Since the West Cumbrian Managing Radioactive Waste Safely Partnership (WCMRWSP) withdrew from the GDF process in 2012, the government has made some changes in legislation that will likely impact the future of nuclear waste. In 2014, the government committed to new nuclear power builds at the same time as classifying a GDF as a major infrastructural project of national import (DECC 2014, p. 5). This prior commitment to new nuclear builds has been reinforced since, with a plan for 24 GW of nuclear energy or 3.5 times the current nuclear capacity (GOV.UK 2024b; Walker & Walker 2022). Most importantly, these changes in policy and nuclear plans ignore the conclusion of the CoRWM that the GDF was for legacy waste and that any newly produced waste needs a new deliberative process and mandate. As well as ignoring this inconvenient conclusion, these changes have had a number of broad effects. Firstly, the inventory of what will go into the GDF has upper and lower ranges presented, depending on the amount of new nuclear capacity achieved (see Table 8.2). Some of the people responding to the WCMRWSP expressed concern that this upper limit might affect the size and process of the GDF after the right of withdrawal has passed (WCMRWSP 2012). There is also a cost implication here with a question of who is paying for this disposal. The various subsidies to the nuclear industry, including loan guarantees, the strike price (minimum price nuclear operators are guaranteed for each unit of energy), and capping of industry costs for waste and any catastrophe, all amount to a large privatisation of profit and loading of costs onto the public (Bridle & Attwood 2016).

More specifically on waste, this can be separated and is financed differently for public and private nuclear waste. The Nuclear Decommissioning Authority (NDA) set up in 2005 and funded by the government took responsibility for BNFL and the Atomic Energy Authority public sector nuclear sites. The NDA in 2013 was spending £2.4 billion annually on nuclear decommissioning, at the time 66% of the then Department of Energy and Climate Change's annual budget (MacKerron 2015). For privately run NPPs, there is another financial arrangement. In 2011, the government capped industry contributions to the waste they produce at approximately £5 billion (DECC 2011). This cap is an estimate of costs from the NDA, but an independent report by Greenpeace (Jackson 2011) suggests that over 40 years per 1.35 GWe PWR, taxpayers will be paying around £427 million or 39% of the overall cost. As shown in Table 8.3, costs have risen such that now the GDF total sits between £18 billion and £53.3 billion. The actual total waste subsidy is unclear, but what is clear is that this subsidy is significant and is a privatisation of profit and socialisation of risk and costs.

Table 8.2 GDF Baseline Costs (2017/18 Monetary Values) and the Uplift in Costs for Uncertainty, Risk, and Optimism Bias

Inventory	*Cost Estimates*	*Lower Technical Complexity*	*Higher Technical Complexity*
Legacy waste (includes wastes already accumulated and waste from the current fleet of nuclear power stations)	Baseline minimum cost estimate	£12.1 billion	£18 billion
	Upper estimate including risk, optimism bias, and uncertainty	£20.3 billion	£32.2 billion
Full potential inventory (includes legacy waste, waste from new nuclear, and nuclear materials that could be categorised as waste in the future, such as spent fuel)	Baseline minimum cost estimate	£20.3 billion	£30.5 billion
	Upper estimate including risk, optimism bias, and uncertainty	£33.1 billion	£53.3 billion

Source: Reproduced and adapted from NWS (2021, p. 27)

A final point on costs is how much of a drain the nuclear is on this country. The NDA and the government report that overall costs of decommissioning sit between £124.4 billion and £263 billion, the upper limit of which is just over half of the government's total on-budget and off-budget liabilities of £514 billion (NDA 2023, pp. 134–135: UK Government Investments 2023, pp. 20–26). This suggests as a nation we are spending somewhere between a third and half of our overall budget on nuclear decommissioning for around 15% of total electricity (which people pay for anyway), medical, industry, and defence uses. Overall, this makes the NWS claims to value and keeping costs low sound a little performative.

Another effect of the changes in legislation is more implicit. This is the possibility that if no community agrees (unlikely but a possibility), then as infrastructure of national import the government might revert to top-down imposition. This would become more of an issue the longer no site is found/agreed, with this potential impasse interacting with increased levels of waste from a programme of new nuclear builds. More broadly, the return to a securitisation discourse and energy policy, as described earlier, has implications for our democratic polity to which we will turn below.

Democratic Deficits and the Nuclear

Presently nuclear waste policy is perhaps the most democratically responsive of the United Kingdom's mega infrastructure projects, or those costing over £1 billion. This was seen earlier in how the Theddlethorpe Community Partnership included in their deliberations a critic of the GDF and nuclear power more generally. These GDF processes can be usefully compared to HS2, which is of similar cost and ostensible national import. The processes involved with HS2 have been far from democratic with compulsory purchases, environmental degradation of nature reserves and ancient woodland, use of private and unaccountable security services, and other top-down injustices (Brock & Goodey 2022). These injustices are compounded by unilateral decisions by the government truncating the whole parts of HS2, meaning that even supporters of this project are now unhappy. The search for the GDF is ongoing and, as stated, a much more deliberative process so far. This remains true despite the problematic subsidisation of the nuclear industry and the associated huge costs of dealing with nuclear waste.

However, it is significant where this deliberation is permitted in the overall technopolitical and energopolitical process. Nuclear waste is a consequence of the civilian and military uses and thus is an end-of-pipe, downstream effect. Deliberating within boundaries and with invites over the refuse of our nuclear energopolitics has been permitted, but only after a long and difficult struggle (Cotton 2017). At the upstream, top-of-pipe, resource or supply end of nuclear energopolitics, there is a distinct lack of democratic engagement and process. There was, as shown earlier, a brief offer of some unspecified consultation that was then quickly buried, and normal upstream energopolitics was resumed. As argued earlier and by the CoRWM, a nuclear power policy that radically expands current nuclear power capacity will have downstream effects. These effects will not just be a higher level of inventory, they will include questions, critique, opposition, and perhaps protest. They will expand the timeline of the GDF, or indeed demand another GDF, adding hundreds of years, or a whole new process, with impacts on people in the near, mid, and long-term future (Thomas 2023). They will also include new risk profiles of both the economic and the technological kind. Economically, there is the risk that taxpayers and energy bill payers become locked in to expensive liabilities and cost inflations, when there are cheaper and viable alternatives (Price et al. 2023; Walker & Walker 2022). Technologically, SMRs have been linked to more fissile and problematic waste that may not be suitable for a GDF (Krall et al. 2021). This sociotechnological risk may be speculative, but then this is usually the case with new technologies, even if based on older technologies. Furthermore, as Collingridge's (1980) dilemma of control framework argued, it is not just new technologies we should be deliberating and discussing. It is also established, entrenched technologies like nuclear that have institutional power and protection, as well as cadres of experts who are incentivised to defend and advocate their cause (Genus & Stirling 2018).

Risk, such as from SMRs and the nuclear more generally, will be assessed by the government and the regulatory regime using formal risk assessment methods. However, these methods are inadequate here. This is because risk is a complex socio-political question that cannot be fully addressed by statistical methods or the market (Stirling 2015). At a rudimentary level, risk can be split into our understanding of the probability of an event and then our understanding of the impact or outcome of the event (Wynne 1992). However, this basic equation of probability times impact is clouded by issues of ambiguity, uncertainty, and ignorance. Ambiguity regards the human and discursive elements of an issue, including contested framings, difficulty in comparing incommensurable objects and processes, and issues of behaviour, trust, interests, and equity. Uncertainty regards the complexity of open systems like the body, society, or the climate; new pathogens; and non-linear processes more generally. Finally, ignorance refers to little or no understanding of probabilities, outcomes, and even the processes or events themselves (Stirling 2007, 2015; O'Riordan 2000; Wynne 1992). This framework of sociotechnical risk can be depicted schematically as shown below:

Table 8.3 Contrasting States of Incomplete Knowledge and Examples

	Knowledge Concerning Probability	*Knowledge Concerning Outcomes*
	Not problematic	*Problematic*
Not problematic	**Risk** • Familiar systems • Controlled conditions • Maximum safe dose of radiation for adults in minority world nuclear sites	**Ambiguity** • Contested framings, questions, assumptions, and methods • Incommensurable comparisons: apples and oranges • Disagreements between specialists and disciplines • Interest, language, and meaning • Issues of ethics and justice
Problematic	**Uncertainty** • Complex, non-linear, open systems • Human element in causal models • Specific effects beyond boundaries • Energy choices to mitigate the CEC • Dangers of nuclear proliferation • Maximum safe dose of radiation for children, pregnant women, and workers and miners of uranium in the majority world	**Ignorance** • Unanticipated effects • Unexpected conditions • Gaps, surprises, and unknowns • Tipping points in climate stability • GDF containment from the socio-environmental over deep time

Source: Adapted from Stirling (2007)

For conditions of uncertainty, we can outline outcomes but have difficulty with assigning probability. We have a good idea of the outcome of overexposure to radiation, partly drawn from some highly unethical experiments on racialised and disabled groups in the United States (Voyles 2015). While probability of risk for adult workers in the more controlled environment of nuclear facilities in the minority world is fairly well established, this is not the case for other groups. One vulnerable group is pregnant women living within 5 kilometres of a nuclear facility and their unborn children. Fairlie (2014) shows there is higher than expected cases of child leukaemia in populations living within 5 kilometres of NPPs in Germany, Switzerland, the United Kingdom, and France. The causal mechanism is unclear; thus, the probability is also hard to establish. Fairlie (2014) speculates it could be down to fuel changes when reactors are opened and depressurised, and there is a significant spike in nuclide emissions. Another exposed group are workers and miners in the majority world's uranium mines. Regulatory agencies and health systems are under-resourced in the majority world, which interacts with the stochasticity of the radioactive alpha particles that lodge into the lung tissue of workers. This can take anything up to 20–30 years to have cancerous effects, which gives mining firms plausible deniability and/or these firms simply disappear due to the boom–bust nature of the industry (Hecht 2012).

With ambiguity, it is not the probability but the outcomes that are problematic. Such ambiguity concerns the connected and interacting social, economic, and environmental impacts. Since the end of the techno-optimistic era in the 1970s, the nuclear seems to have existed in a general state of ambiguity. The framing of it as a solution or the problem is ongoing as is evident in the latest UK public attitudes on energy sources. Nuclear power is seen as a reliable source of energy, with 50% agreeing and 9% disagreeing, and similar proportions think nuclear power could help mitigate the CEC, with 44% agreeing and 10% disagreeing. However, this positivity falls concerning nuclear power's affordability, with 38% agreeing and 10% disagreeing; then considering nuclear power's safety, there is an even more significant drop in confidence, with 34% agreeing and 20% disagreeing (Desnz 2023). In contrast, support/strong support for renewables is much higher at 82%, with only 2% opposing or strongly opposing this source of energy (Deznz 2023). Ambiguity has also been evident in the search for the GDF, with a rather bitter dispute between two geologists during the previous site investigation in Cumbria over the suitability of the host geology (Cotton 2017). There was also a committee within the CoRWM deliberative processes that was described as dysfunctional, resulting in one committee member's acrimonious dismissal and another resigning (Chilvers & Burgess 2008). The central dispute was over the role of expertise, with nuclear scientists claiming the social scientists were trying to "reject[s] expert input in favour of public consultation" and that as "rationalists" they could not accept this "perversion of reality" (Baverstock & Ball 2005, pp. 316–320). Thus, we have two clear instances of nuclear ambiguity within and across disciplines. Additionally, in the latter example, we have a quite scientistic position from nuclear scientists that was published in a nuclear journal. This also exemplifies the ambiguous role of interest groups, and technology experts and their institutions acting in biased and defensive modes (Collingridge 1980; Genus & Stirling 2018).

Finally, we have ignorance where we have no or little understanding of either probability or possible impacts and outcomes (Stirling 2007). Many environmental impacts and processes remain in this space, such as the broad effects and impacts of the CEC over time, and the so-called tipping points within this process, or the non-linear feedback mechanisms that exacerbate the overall crisis and instability (Lenton et al. 2008). The containment of the radioactive material in the GDF over 200,000 years also falls into this space of ignorance, as controlling for interacting socio-environmental systems over deep time is extremely difficult (Ramana 2020).

Thus, certain aspects of the nuclear can be dealt with using formal risk analysis and statistical methods. However, the uncertain exposure of specific populations to radiation; nuclear power's ambiguous social, economic, and political effects; and the unknowns and ignorance about the GDF over deep time require more than these narrow methods. Broader deliberative methods are the case with the search for a GDF, and this is laudable, even if there is room for improvement – such as greater involvement of young people in this deliberation. But this deliberation needs to be moved upstream to our energy choices. This is because risks cannot be completely removed. Perhaps not using nuclear power would present other risks, such as those connected to relying on carbon capture and removal technologies. This is the nature of the uncertainty surrounding dealing with the CEC, and these are ultimately political and social questions with technical details. As Stirling (2007) and the CoRWM said, science here should be on tap not on top. These are monumental decisions about our collective future. As such, these are questions we the public should have a say in, especially the younger part of the public.

Conclusion

This chapter examined the United Kingdom's historical nuclear power and waste processes and current policies. Historically, nuclear in the United Kingdom has involved aspects of nuclear colonialism in the testing of nuclear weapons on colonised land and questionable technology in terms of reliability and cost. The overall governance of the nuclear was top-down, secretive, and undemocratic, while the waste policy was at best an afterthought. This history is at odds with both the policy-maker and the NWS framing of the nuclear as historically providing "great benefits." Perhaps it did provide great benefits to specific groups in specific places and positions in the military industrial complex, and in a more diffuse way to the urban core over rural peripheries. However, benefits in one place and time at the expense of another place and time is not justice but extraction. Also, the overall costs of the nuclear, around half of our on- and off-budget expenses and liabilities as a nation, call into question even the utilitarian calculus of the greater good to the majority. Largely because of the failures and problems of this top-down nuclear governance, nuclear waste policy is, comparative to other mega projects, a democratic process with many positive aspects. However, this is undermined by the wider nuclear policy, its huge costs to the public, and the lack of democratic process and discussion about these costs and broader energy choices.

We call for a new round of deliberation mirroring the scale and scope to the CoRWM process, but this time democratically and deliberatively engaging and discussing our energy choices and future. This process should include children and young people as it is their future we are deciding on, and as argued in Chapter four, these young people can be considered proxies for unborn future generations. This deliberative process should be ongoing as the intergenerational democracy model advocates. This is because the CEC will be an ongoing process that we must adapt to over time. Also, we need to democratise our chronopolitics, our technopolitics, and our energopolitics – to monitor not only new technologies but also established ones (Collingridge 1980). This would be best done with ongoing, transparent, and inclusive deliberation and more distributed forms of energopower. If nuclear power is needed, its advocates can make their case in this forum, as can those advocating for alternatives. This is how democracy arguably should work, and under the stress of the CEC might have to work (Dryzek & Pickering 2019). Failure to offer these democratic choices over our future risks a deficit, and this would not be of the public understanding, but rather on the nuclear as a democratic deficit internal to its energopower model.

Note

1 All subsequent citations are abbreviated to WG (working group + n of meeting) or CP (Community Partnership + n of meeting). The minutes are also all available on the webpage provided in the references.

Conclusion

Justice for All

This book explored the literature and ideas of intergenerational justice, endeavouring to show how this should work with, not against, intragenerational justice, and then looked at case studies of nuclear waste management. The main argument made is that liberal philosophy and its empirical tool mainstream economics make justice scarce. They have led us to the CEC and offer few credible solutions, just more of the same with tweaking. Thus, we argued we need new thinking, new actors, and new methods and practices to get us out of these crises. The new thinking is relational, seeing humans as connected to each other and the web of life. It is informed by indigenous perspectives but is not extracting knowledge as it seeks redress, land back, and justice for all. And it sees time as both linear and spiral and understands we must look back to move forward. The new actors are the young, the marginalised, and the people most affected by developments. Finally, the new methods involve reflexive deliberative democracy that understands positional power and works to mitigate these dynamics to create desirable deliberative spaces. It is time for deliberative decision-making and deliberative exercise for all.

This conclusion will consider nuclear waste management and governance through the four environmental justice tenets of distribution, process, recognition, and redress. It will then compare indigenous philosophy/practice to Western science via the idea of haunting and ghosts. Finally, it will offer conclusions on how to build a just future drawn and synthesised from the literature that this book stands upon.

Nuclear Waste Management and Justice

Nuclear waste management, akin to the CEC, is an issue of both intra- and intergenerational justice. This is because both involve asymmetric distributional impacts, or injustices, laid upon peripheralised people across time and space. For instance, as we have seen in New Mexico, marginalised groups face the impacts of "cradle to grave" nuclear degradation, while the energy produced serves urban centres and profits of both civilian and military uses accrue to military industrial complex actors. Looking to the future, our legacy

DOI: 10.4324/9781003422686-12

waste certainly needs to be dealt with, but this does not remove the fact that future generations will not have used any of this energy, nor are they likely to "benefit" from the increasingly dangerous geopolitical situation that has the Doomsday Clock closer to midnight than ever before (The Bulletin of Atomic Scientists 2024). Moreover, as our leaders increasingly use bellicose language, militaristic actions, and nuclear weapons policy (Norton-Taylor 2024; Young 2024; Hudson 2024), they fail to do what is required to mitigate GHG emissions to stabilise the climate for present or future generations (Stoddard et al. 2023; Anderson 2023a).

In the three cases of nuclear waste management, we found differing problems and issues. The unifying theme of these problems was procedural democracy, with one of two areas deliberating over a GDF in Canada found wanting, the more general process in Finland being top-down, and the United Kingdom's search for a GDF being hampered by the government's nuclear ambitions, its inability to capture an appropriate scale of engagement, and its lack of learning from the past. All cases presented issues in regard of intergenerational democracy, with children and young people only involved in a largely instrumental sense, either to recruit these young people into the nuclear sector or to manage public relations. Thus, while the search for a GDF in England, and more broadly internationally, is on the more deliberative end of mega-projects costing more than £1 billion (Cotton 2017), there is room for improvement and learning from the past. This learning and improvement will be dealt with in four parts following environmental and energy justice ideas around the justice of distribution, process, recognition, and reparation/restoration (Van Uffelen et al. 2024).

Distributional Justice

As argued throughout this book, liberal representative democracy and its utilitarian equations mean peripheralised communities and more-than-human nature are largely excluded from the "greater good." Furthermore, under neoliberal governance, this "greater good" seems to be benefiting fewer and fewer more and more. As a brief example of this, UNICEF (2023) found that the United Kingdom has presided over a staggering 20% increase in the rate of child poverty since 2014. This is at the same time as successive governments have bailed out bankers following 2008, and then the wealthy more generally during the pandemic via quantitative easing. This "socialism for the rich" reached a massive £2,285 billion in 2020 alone (Statista 2023). This upward extraction has resulted in a record amount of new UK billionaires and their share of wealth, growing from 53.9 billion in 1990 to 653.1 billion in 2022, equating to a 1,000% increase (The Equality Trust 2022). Of course, the NWS is not the government and is limited in what it can do in terms of redistributing wealth, apart from perhaps making sure the income ratios between junior and senior members of the staff/board do not become too excessive. This is something the NDA is signally failing to do (NDA 2023, pp. 101–102). Nevertheless, there is one thing that the NWS could offer to host communities that would act to redistribute wealth and could be seen to partially address the distributional aspects of justice.

This would be the offer of a community energy – wind, solar, or hydro – that could partially power these households and/or pay a yearly dividend to the host community. In a basic sense, this inverts the utilitarian equation of the nuclear where the small group takes on the cost/illth for the wider society. Community energy benefits the community that owns and manages the infrastructure first and foremost; then there is a diffuse benefit that accrues to the rest of the society by contributing to decarbonisation. In terms of knowledge, the community could learn about energy and how it is planned, produced, and distributed; its costs and benefits; and much more. After all, we do need communities on board with the energy transition, and the best way to engage them is involving them in the processes and outcomes that community energy involves (Berka & Creamer 2018). Temporally, while renewable infrastructure does have projected lifetimes like any energy infrastructure, renewable infrastructure is future facing and thus could by proxy associate NWMOs with a desirable future. Furthermore, every act of reducing our environmental impact benefits future generations. Community energy has long been associated with action on energy poverty, thus supporting and offering an ethic of care to some of the most marginalised people in our society (Nolden et al. 2022). More generally, smaller-scale renewable infrastructure has less impact on wildlife and ecosystems; therefore, it has the potential to move the subject of justice beyond the human and the anthropocentric. Finally, in terms of scale, in a similar way to inverting the utilitarian equation, these peripheralised communities that have often hosted national-scale infrastructure, and potentially will with a GDF, would now benefit from a local-scale form of energy infrastructure that they could relate to more positively.

Procedural Justice

Our representative system can be characterised as a very thin form of democracy with very little involvement of the subjects of this system, with even deliberative democracy reflecting this problem. Clouding the discourse in this system is the mainstream media and various corporate interests that play an outsized and undemocratic role in gatekeeping the common sense of possibility and maintaining vested interests (Herman & Chomsky 2002; Anderson 2023b). Thus, it is a system riddled with procedural injustices and dysfunctions. However, deliberative democratic endeavours to re-democratise at the local level run into issues due to a poor conceptualisation of social positional power. This involves an outsized role given to experts, as was the case with nuclear waste deliberations in the United Kingdom in the 2000s (Chilvers & Burgess 2008). This is not to say experts should not be involved, but ethical questions should we do this? why should we do this? who/what does this affect and how? – are central to wicked socio-environmental issues and should not involve decision-making hierarchies. The intergenerational justice model is quite clear on this, with the non-experts controlling/overseeing the overall process and experts and lay people equal on the decision-making level. Another aspect of this is the use of random selection of participants that, like jury service, tends to exclude all kinds of people, like those in temporary housing, students, homeless people, and more generally those who cannot afford the time/expense of taking part. A further, less tangible, but no less important aspect of this is the way deliberative discourse is done. This favours those educated to higher levels and able to structure arguments logically and express these calmly (Young 1990). Overall, these aspects of power make deliberative spaces quite hostile and exclusionary to many types of people. Thus, we need to revitalise and rethink our democracy at multiple levels.

More specifically, following the intergenerational democracy method, the NWS could institute a new round of deliberation. This method involves three main advantages to more traditional forms of deliberative democracy, and that would address some of the problems with these traditional forms. Firstly, it should use non-random methods of selection to get as close and accurate a cross-section of society – including from age 9 to beyond 80. Those with low incomes should be reimbursed for their time as this factor often excludes these groups. Secondly, it should be layperson-led with the intergenerational reference group made up of diverse members of the community – not controlled and directed by experts. Discursively, there should be an allowance for different modes of discussion, those of greeting, rhetoric, and storytelling (Young 1990), and an acceptance of the central role emotions play in human cognition (Taebi et al. 2021). This acceptance of the central role of emotions might allow for a movement away from the patriarchal and rationalist model of discourse. Obviously, experts would be there to advise and facilitate, but they should not act as decision-makers in this method. This is not to say that all are deliberate on the technical aspects of a GDF – this would clearly not work and the role of experts here is to be transparent and communicate risks and issues in an accessible but non-patronising way. But we should all be deliberating on issues that directly impact us. This includes deliberation over *more nuclear power* as this is not a technical question but a socio-political and ethical one about the kind of future we want. Finally, this intergenerational democracy method should be ongoing with milestones and evaluation meetings – in effect trying to make it a kind of longitudinal method but directed by the participants. This would correspond with the timeline of the GDF and act against the increasing attenuation of our democracy (Cesnulaityte et al. 2022; Dryzek 2000, 2013).

The Justice of Recognition

Recognition justice goes beyond tolerance of difference, stipulating all should enjoy full, equal rights (Jenkins et al. 2016; McCauley et al. 2013). It demands that people must be free from physical threat, disrespect, insults, and degradation and other forms of misrecognition, including non-recognition and misrepresentation, which can result from cultural and political domination (Schlosberg 2004; Heffron et al. 2015). Clearly, the use of the deficit model of public understanding, which, as seen in part two, is common in the nuclear space, is a form of misrecognition. There is no ignorant mass of "public" out there – there is in fact a highly diverse and fractured/networked society diverse in knowledge, ideologies, and subjectivities. However, the number of times the authors have heard in nuclear spaces forms of this deficit model suggests it is a form of sticky knowledge (Cotton 2017). One example involved a senior nuclear stakeholder saying that speaking to the public was like speaking to children. Thus, there is this general problem. More specifically, there are three groups that are either not recognised or misrecognised in nuclear spaces, and these are children and young people today, the more-than-human, and future generations. This book has advocated for intergenerational democracy and argued against anthropocentric thinking which treats the world as a resource, an endless series of inputs and outputs. Here we shall think a little more about the difficulty of recognising the unrecognisable – unborn future generations.

Future generations, as we saw in the intergenerational justice literature, are generally discriminated against by our system, which discounts them in various ways. In the nuclear space, these are generally talked about as having a direct interest in nuclear waste solutions, such as a GDF. This makes sense as it is dangerous to leave this waste where it is, and waiting for a tech-solution that may never work or appear seems ill advised. However, when, as in the United Kingdom, this nuclear waste solution becomes complicated by plans to significantly expand the nuclear power capacity and therefore the GDF inventory, this idea of a shared interest between future generations and the present becomes stretched and tenuous (Blowers 2020). There is a fundamental difference between legacy waste, which was poorly understood and one of the inevitable unintended consequences of a technology, and deliberately creating more of this waste when it is known how dangerous, long-term, and difficult and expensive it is to deal with. We cannot know what future generations would prefer, but we could expect that they would not choose a society that routinely shifts burdens onto mis/unrecognised and marginalised sections of this society, including future generations. Or maybe they would, *if this was done to them.* Perhaps then the question is, should we as the present generation break this vicious cycle?

The Justice of Redress and Reparation

Olúfẹ́mi Táíwò (2022), in his book *Reconsidering Reparations*, argues we must look to the past and how historical development accumulates both wealth and illth under capitalism. In contrast, using inequality as a metric takes a snapshot in time, thereby occluding how colonial history has produced today's billionaires and economic empires and their opposite yet foundation – the billions who lack the basics of social reproduction. Thus, Táíwò (2022) argues we need to intervene now to redistribute resources and power to reverse the upward transfer of wealth and to improve lives and intragenerational justice. Also, this would push this redistribution forward to future generations and begin the long road towards intergenerational justice and a just world (Táíwò 2022). In a similar vein, Astra Taylor (2021) uses the concept of security and its opposite insecurity, pointing out the latter term only emerges in the 17th century and is a product of the rise of capitalism. Taylor accepts that humans have always had a more ontological precarity – we get ill, have accidents, and, as Nussbaum (2004) argues, are fundamentally vulnerable over time. The term security comes from the Latin *securitas* – free from worry and from *sine cura* – without care, suggesting humans have always been concerned with this fundamental vulnerability and how to insure against it. However, our system, as well as producing commodities and widgets, also systematically produces economic insecurity (Taylor 2021; Mau 2023; Dorling 2023), thus exacerbating our built-in vulnerability. This has an important consequence relevant to intra- and intergenerational justice and the climate crisis.

This insecurity or constructed scarcity is blocking effective climate action. The UK's new government, the Labour Party, had prior to winning the general election already been managing expectations by watering down its already insufficient climate mitigation funding (Stacey & Crerar 2024). This was insufficient because £28 billion per year would not even cover the essential national housing retrofit, incidentally the likely main victim of this policy U-turn (Stacey 2024). This retrofit would cost around £100.15 billion over three years (probably more now considering inflation) but would arguably pay for itself over a 25-year period (Brown et al. 2020), *as it improved lives*. The same Labour Party has also recently promised it will not limit bankers' remuneration packages and instead will "champion of a thriving financial services industry" (Simpson et al. 2024). This thriving industry, however, is at the heart of a global tax evasion network, as is the United Kingdom more generally (Christophers 2020). This means the financial service industry is contributing to the financial scarcity the Labour Party is citing to justify the climate mitigation U-turn. Also, this industry is still funding fossil fuel expansion (BankTrack 2024), as is the United Kingdom more generally, as such exacerbating the climate crisis. Thus, within our political elite and in mainstream discourse more generally, we have a binary – economy/environment, with the economy nearly always winning. This socially constructed binary is exploited by powerful people – like bankers and fossil capital – to prevent, slow, and shape climate action to their own ends (Kenner 2019; Anderson 2023b). Overall, there is an irony here as no one will be secure under climate breakdown as our economy sits within the environment (Raworth 2022). However, those with the least responsibility are now facing and will in the future face the worst of this crisis. This is wrong and should be redressed, and this redress begins with radical redistribution of wealth and power.

What is role for the nuclear in this redress? Firstly, both the United Kingdom and United States have summarily used and appropriated First Nation societies' land for nuclear weapons development, testing, and nuclear waste processing (Voyles 2015; Gómez 2022; Alexis-Martin et al. 2021). These are ongoing injustices, and if these states and the nuclear industry wish to be part of a just transition, then they need to move towards reparations and redress for these communities. Historical wrongs have contemporary effects, especially regarding land, and should not be simply ignored or just rhetorically acknowledged – there should be some form of material and social redress. More specifically for the NWMOs, the community benefits need to be substantive, not filling gaps in the finances of peripheralised areas, as in Finland's case. The example of community renewable energy given earlier is one way, but importantly these benefits should not merely replace something withdrawn or that communities should expect in a highly developed economy. This could look like a bribe or what Lehtonen and Kojo (2019, p. 196) called fatalistic "bought voluntarism."

Finally, we need transparency about nuclear finances, the subsidies, and how much taxpayers are really paying for this industry. When we are told we cannot afford to rapidly decarbonise by both major parties, we should be aware of where our money is going, why, and if it is worth it. Are we using an energy technology that is far more expensive than renewables regardless of nuclear waste/decommissioning costs (Lazard 2020), to covertly subsidise nuclear weapons capability (Stirling & Johnstone 2018; Johnstone & Stirling 2021)? Transparency around civil and military nuclear finances is far from the case now (Bridle & Attwood 2016; Norton-Taylor 2024), as shown in Chapter eight, and this should change. The next section will move on to the more philosophical questions that we considered in this book.

Ghosts of Seppo and Western Science

This section will consider the fascinating anthropological study by Ialenti (2020a) of the late head of the Finnish nuclear authority safety case – Seppo – whose influence lives on as a kind of haunting presence. This example will then be used to speculate on how Western science and its natives (Sahlins 2017) may be haunted by their ghosts of past, influential, mainly White men, of science. This may explain why similar problems such as the deficit model, despite being soundly deconstructed and critiqued, keeps re-emerging as seen in part two of this book and reported in the literature (Simis et al. 2016; Suldovsky 2016; Sturgis & Allum 2004). It will then compare animist/indigenous treatments of ghosts of the past and speculate on whether an open acknowledgement of Western ghosts of scientific rationality could help lay them to rest. This may also allow us to surpass scientistic rationalities that misrecognise not only people but also science itself.

Seppo was the *de facto* authority of nuclear waste disposal in the 1990s in Finland. Ialenti (2020a, p. 6) describes how the Finnish safety case emerged "from wide assemblages of geographically distributed expert communities, international institutions, industry initiatives, government programmes, university researchers, and legal entities." However, in mastering this specialist and esoteric knowledge, one could become indispensable to an employer, as Seppo did, and if this person was to leave, or in Seppo's case unexpectedly die, work on the safety case would and did temporarily grind to a halt. By 2000, Seppo was untouchable as the titular safety case leader. However, when he died in 2005, Posiva was forced into speeding up a process it had been more slowly moving towards, but which Seppo had been resisting, of decentralising the safety case team into ten experts (Ialenti 2020a). However, it is Seppo's afterlife that concerns us here.

A first issue Ialenti (2020a) points out is that as a leader Seppo was a difficult figure, with the flip side of his indispensability being rudeness, volatility, and aggression. Those who follow in his footsteps seemed to split, with some using him as an example not to follow as poor interpersonal demeanour is neither good for team dynamics nor acceptable in workplace ethics. Others valorised the man as committed and passionate, with one informant claiming that while Seppo may have been morally ambiguous, he was "usually right," while another claimed "to be careful was his hobby" (Ialenti 2020a, p. 8). A second issue was Seppo's ongoing influence, best seen in his right-hand man Gustav, still working for Posiva during this study, who Seppo fired twice only to rehire him when he had regained his temper. Seppo, for Gustav, was also an ambivalent figure, both "Stalin" and a respected and missed colleague, but more than this, Seppo was a haunting figure that influenced Gustav and his work and views beyond the grave in both legitimate and technocratic ways. On one level, Seppo had a dim view of the increased bureaucratic nature of the safety case, including more pointless documentation – something many in education and business will recognise as the bullshitisation of work (Graeber 2018). However, on another level, Seppo was deeply technocratic and believed the nuclear waste problem was solved and that the only task left was to produce reports that spelled this out for the less intelligent. These were sentiments Gustav repeated in his critique of the safety case, which he saw as increasingly "bloated, costly" and undermined by excessive documentation (Ialenti 2020a, p. 11).

As Ialenti (2020b) makes clear, this was during a period of neoliberal restructuring of Finland's research institutions. This involved state funding replaced by private funding, commercial and private interests gaining more influence over direction, alongside the casualisation of research and university positions. Thus, we can see a number of things going on here – a situation getting worse but also in many ways improving, but this being all dismissed as one thing. This frames the solution as going back to the good old days when we could rely on a few experts. However, we would argue the neoliberalisation and bullshitisation of jobs, and all that this entails of big businesses shaping research agendas, the casualisation and precarity of university and research jobs, are one aspect and one that should be resisted. In contrast, the use of precaution and the precautionary principle of testing and testing again and thinking from different disciplines and perspectives, in effect seeing the wicked problem as just that, is a good thing and needs to be built on and expanded.

Returning to the influence of Seppo, Ialenti (2020a, p. 14) claims "*expert afterlives* to be temporal, epistemic and affective and shape the technocratic imaginations, projects and knowledge production." This is perhaps inevitable in fields with highly specialised knowledge like the nuclear sector, and Ialenti (2020a, p. 14) concludes more generally that it would be fair to surmise that figures like Seppo are soon to be or already haunting "nuclear experts elsewhere." To be fair to the nuclear scientist, we would extend this to mainstream positivist science more generally and speculate that the zombie-like return of the deficit model of public understanding is perhaps partially due to the voices of the dead, *great* men, from Descartes to Hawking, to the more local avatars, all preaching a form of reductive scientism that cannot deal with wicked problems because they are fundamentally systemic and relational problems not amenable to reductive Cartesian thinking. Moreover, this scientistic thinking appears to misunderstand what science is and what it is good at. Science is, if it can be nailed down to a unified thing and this is questionable (Erickson 2016), a way of understanding and acting on the world. However, in acting on the world, it changes the world and shapes people and things in it. This is what Haraway (1988) and Habermas (1972) mean when they say we discover and create the world. A good example is evolutionary science. For much of its history as it was done by privileged White men, it reproduced a story about the world that focused on competition for resources and for mates that was a good fit for heteropatriarchal, racialised capitalism (Harris 2023; Rushcoff 2022; Wynter 2003). However, more recently as women and more ethnically diverse evolutionary scientists have emerged, we get a more complicated picture (Hustak & Myers 2012), comparable to the one Graeber and Wengrow (2021) offer for the long human history. This is a kaleidoscopic mix of cooperation and competition but more – a fundamental relationality and interdependency that join in questioning the anthropocentrism and individualism of our system. This is because from our bodies to and through the flora and fauna, there is a web of life (Haraway 2016; Kimmerer 2013). Thus, science is good at describing and measuring things, with the caveat that sometimes describing and measuring things have unexpected effects and are perhaps counterproductive if not done reflectively and contingently. After all, science is supposed to be contingent, hence the need to test hypotheses. What science is not good at is telling us what we *should* do in the world, and how to formulate ethics and deal with complex socioecological problems like the CEC, as these are fundamentally moral and political questions. As such, these problems should be deliberatively dealt with by the communities affected, including the scientists who should facilitate and provide balanced information. However, in decision-making stages, these experts should be no more and no less than their fellow lay citizens. As argued earlier, ethical expertise is a contradiction in terms, as ethics is morality considered and reflected upon, and everyone can and should be doing this.

Another way of understanding and acting on the world are the animist and indigenous traditions and philosophies. While these traditions are many and distinct, there are commonalities that can, as Kimmerer (2013) does, work with sciences and complement them. Briefly, the commonalities include a relational worldview rather than binary thinking and the overarching division of society/nature. This relational worldview avoids anthropocentric and individualist thinking. Time and history are not seen as linear but more of a spiral or cycle like day and night or the seasons. One consequence of this is figures from the past are openly acknowledged and consequential with ancestors and history "memorialised in the landscape" and part of an intergenerational bridge connecting past to present and to the future (Winter 2022; Estes 2019; Sahlins 2017). The fact that the IPCC (2022) points towards indigenous knowledge/philosophy as a solution to the CEC is powerful institutional recognition of these ways of thinking. However, we cannot culturally appropriate this way of thinking without doing the redress. This means the ending of settler colonialism and returning land, rights, and justice of distribution, process, and recognition to the native peoples in the settler colonial states. It also demands global justice and the ending of the minority world's predatory extraction from the majority world. Finally, it requires both the end of discrimination against capitalism's many others and the ending of our internal colonisation – the voice that tells us this is the best we can do, who we are, and what we deserve. A vast machinery of consent manufacturing and/or open violence arrayed against a long history of resistance to this conditioning shows us otherwise.

The Darkness of the Grave or the Womb?

Humanity stands at a crossroads or inflection point. Going forward with the same tools and ways of thinking that have led to this point seems at best unjust, and at worst, suicidal. However, this should not be seen as a bad thing for we live in a world in which even those in the 1% and upwards are not happy, those in the middle suffer from eco-anxiety and guilt due to carbon lock-in and a sense of powerlessness, while the rest are forced to scramble to achieve basic needs and given little time or space to imagine a better world. Again, is this the best we can do? This book has identified liberal philosophy for the parochial Eurocentric discourse that it is and mainstream science as its empirical emanation, moving from *I think, therefore I am*, to *I measure, therefore it is thus*. This way of thinking achieved wonderful and terrible things, but the wonderful was always reliant on the terrible, and the road is running out anyway. Perhaps a better question than can we uncouple economic growth from CO_2 emissions and the broader CEC is can we uncouple the wonderful from the terrible?

We believe yes, and this might involve a number of connected praxes. The first would be a commitment of justice for all, in the sense of capabilities so that all have the space, respect, and resources to flourish. This, as argued in part one of this book, would require minimums and maximums of resources for we live in a finite world, as science should help us understand. The second would be the explicit connection of intragenerational justice to intergenerational justice, for justice in one place/time at the expense of another is not justice but extraction. The third would be a drive to re-democratise our societies, with the cognition that representative democracy is not some end state of social evolution but a particular historical formation and compromise that was largely designed to limit popular power. Economically, most of us still live and work in a variety of top-down dictatorships, which is to say democracy without economic democracy is thin gruel. Thus, a bottom-up, socio-economic democracy could and should be something to strive for. The fourth would be a reorientation of our economies away from conspicuous consumption, prison, and military–industrial complexes towards care and caring institutions, which would have the added benefit that this lines up more with steady-state or degrowth economics. Finally, it would involve understanding our place in the world, as entangled, acting on, and being acted on, part of the web of life.

How do we get there might be the most fraught question. However, the answer is we are close in multiple places now and have been throughout human history. Kim Stanley Robinson (2020, pp. 425–428) offers this answer for the present in his *Ministry for the Future*. In over three pages, he lists cooperatives, voluntary organisations, environmental NGOs, and businesses that are already practising one or more of the above. Graeber and Wengrow (2021) similarly show this in multiple places and times in the past. Finally, as Winter (2022) argues, indigenous societies, in all their variety and despite terrible historical wrongs and ongoing oppression, currently in many places, live and act in accordance with inter-/intragenerational justice and the understanding that we are a part of and interdependent on the web of life. Perhaps the work is to understand these examples, connect them, and build on them. Then maybe this darkness going forward might be, as Rebecca Solnit (2004) speculated, the darkness of the womb and not the grave.

References

Adam, B., & Groves, C. 2007. *Future Matters*: Action, *Knowledge, Ethics*. Leiden, Brill.

Adams, M. 2016. *Ecological Crisis, Sustainability and the Psychosocial Subject*. Basingstoke, Palgrave Macmillan.

Aldeia, J., & Alves, F. 2019. Against the environment. Problems in society/nature relations. *Frontiers in Sociology*, 4(29).

Alexis-Martin, B., Bolton, M. B., Hawkins, D., Tisch, S., & Mangioni, T. L. 2021. Addressing the humanitarian and environmental consequences of atmospheric nuclear weapon tests: A case study of UK and US test programs at Kiritimati (Christmas) and Malden Islands, Republic of Kiribati. *Global Policy*, 12(1), 106–121.

Alexis-Martin, B., & Davies, T. 2017. Towards nuclear geography: Zones, bodies, and communities. *Geography Compass*, 11, e12325.

Alhakami, A. S., & Slovic, P. 1994. A psychological study of the inverse relationship between perceived risk and perceived benefit. *Risk Analysis*, 14, 1085–1096.

Anderson, K. 2023a. IPCC's conservative nature masks true scale of action needed to avert catastrophic climate change. *The Conversation*. [Online]. [Accessed September 2023]. https://theconversation.com/ipccs-conservative-nature-masks-true-scale-of-action-needed-to-avert-catastrophic-climate-change-202287

Anderson, K. 2023b. A habitable Earth can no longer afford the rich – and that could mean me and you. *Climate Uncensored*. [Online]. [Accessed December 2023]. https://climateuncensored.com/1485-2/

Appadurai, A. 2013. *The Future as Cultural Fact: Essays on the Global Condition*. London, Verso.

Arent, D. J., Tol, R. S. J., Faust, E., Hella, J. P., Kumar, S., Strzepek, K. M., Tóth, F. L., & Yan, D. 2014. Key economic sectors and services. In: Field, C. B., Barros, V. R., Dokken, D. J., et al. (eds.) *Climate Change 2014: Impacts, Adaptation, and Vulnerability. Part A: Global and Sectoral Aspects. Contribution of Working Group II to the Fifth Assessment Report of the Intergovernmental Panel on Climate Change* (pp. 659–708). Cambridge, Cambridge University Press.

Arnold, L. 1992. *Windscale 1957: Anatomy of a Nuclear Accident*. Basingstoke, Palgrave Macmillan.

Aronoff, K. 2021. *Overheated: How Capitalism Broke the Planet – and How We Fight Back*. London, Bold Type Books.

Arrondel, L., & Masson, A. 2006. Altruism, exchange or indirect reciprocity: What do the data on family transfers show? In: Kolm, S. C., & Ythier, J. M. (eds.) *Handbook of the Economics of Giving, Altruism and Reciprocity: Applications Volume 2*. Utrecht, Elsevier.

Asafu-Adjaye, J., Blomquist, L., Brand, S., Brook, B., DeFries, R., Ellis, E., Foreman, C., Keith, D., Lewis, M., Lynas, M., & Nordhaus, T. 2015. *An Ecomodernist Manifesto*. [Online]. [Accessed September 2023]. http://www.ecomodernism.org/

Banktrack. 2024. *Banks*. [Online]. [Accessed December 2023]. https://www.banktrack.org/banks#country=United%20Kingdom

Barthes, R. 1977. The death of the author. In: Barthes, R. (ed.) *Image, Music, Text* (pp. 142–148). London, Fontana.

Baverstock, K., & Ball, D. J. 2005. The UK committee on radioactive waste management. *Journal of Radiological Protection*, 25(3), 313.

Baylis, J., & Stoddart, K. 2012. The British nuclear experience: The role of ideas and beliefs (part one). *Diplomacy & Statecraft*, 23(2), 331–346.

Bebchuk, L. A., & Hirst, S. 2019. *Index Funds and the Future of Corporate Governance: Theory, Evidence, and Policy* (No. w26543). Cambridge, MA, National Bureau of Economic Research.

Becker, G. 1988, March. Family economics and macro behaviour. *American Economic Review*, 1–13.

Beckerman, W., & Hepburn, C. 2007. Ethics of the discount rate in the stern review on the economics of climate change. *World Economics*, 8, 187–210.

Beckerman, W., & Pasek, J. 2001. *Justice, Posterity and the Environment*. Oxford, Oxford University Press.

Beddows, D. C., & Harrison, R. M. 2021. PM10 and PM2. 5 emission factors for non-exhaust particles from road vehicles: Dependence upon vehicle mass and implications for battery electric vehicles. *Atmospheric Environment*, 244, 117886.

BEIS (Department of Business, Energy and Industrial Strategy). 2018. *Implementing Geological Disposal – Working with Communities: An Updated Framework for the Long-Term Management of Higher Activity Radioactive Waste*. [Online]. [Accessed September 2023]. https://assets.publishing.service.gov.uk/media/65a7e79fb2f3c60013e5d451/implementing-geological-disposal-working-with-communities.pdf

Bell, M. 2021. Spatialising procedural justice: Fairness and local knowledge mobilisation in nuclear waste siting. *Local Environment*, 26, 165–180.

Bell, M. 2022. The epistemic tensions of nuclear waste siting in a nuclear landscape. *Nature and Space*, 0, 1–22.

Benjamin, W. 1968. *Illuminations: Essays and Reflections*. New York, Schocken Books.

Bergmann, Z., & Ossenwaarde, R. 2020. Youth climate activists meet environmental governance: Ageist depictions of the FFF movement and Greta Thunberg in German newspaper coverage. *Journal of Multicultural Discourses*, 15, 267–290.

Berka, A. L., & Creamer, E. 2018. Taking stock of the local impacts of community owned renewable energy: A review and research agenda. *Renewable and Sustainable Energy Reviews*, 82, 3400–3419.

Berkhout, F. 1991. *Radioactive Waste: Politics and Technology.* Abingdon, Routledge.

Berlin, I. 1957. Two concepts of liberty. In: Hardy, H. (ed.) *Liberty* (pp. 166–217). Oxford, Oxford University Press.

Bhaskar, R. 1997. On the ontological status of ideas. *Journal for the Theory of Social Behaviour*, 27, 139–147.

Bickerstaff, K. 2012. "Because we've got history here": Nuclear waste, cooperative siting, and the relational geography of a complex issue. *Environment and Planning A*, 44, 2611–2628.

Bird-David, N., & Darr, A. 2009. Commodity, gift and mass-gift: On gift–commodity hybrids in advanced mass consumption cultures. *Economy and Society*, 38(2), 304–325.

Blau, P. M. 1964. Justice in social exchange. *Sociological Inquiry*, 34(2), 193–206.

Blowers, A. 2010. Why dump on us? Power, pragmatism and the periphery in the siting of new nuclear reactors in the UK. *Journal of Integrative Environmental Sciences*, 7, 157–173.

Blowers, A. 2020. The legacy of nuclear power and what should be done about it: Peripheral communities and the management of the nuclear legacy. In: Brunnengräber, A., & Di Nucci, M. R. (eds.) *Conflicts, Participation and Acceptability in Nuclear Waste Governance: An International Comparison Volume III.* Heidelberg, Springer VS.

Blowers, A., & Leroy, P. 1994. Power, politics and environmental inequality: A theoretical and empirical analysis of the process of peripheralisation. *Environmental Politics*, 3, 197–228.

Boehm, C. 1999. *Hierarchy in the Forest: The Evolution of Egalitarian Behaviour.* Cambridge, MA, Harvard University Press.

Bonneuil, C., & Fressoz, J. B. 2016. *The Shock of the Anthropocene: The Earth, History and Us*. London, Verso.

Bridle, R., & Attwood, C. 2016. *It's Official: The United Kingdom Is to Subsidize Nuclear Power, but at What Cost? Global Subsidies Initiative*. International Institute for Sustainable Development. [Online]. [Accessed September 2023]. https://www.iisd.org/publications/report/its-official-united-kingdom-subsidize-nuclear-power-what-cost

Brock, A., & Goodey, J. 2022. Policing the high speed 2 (HS2) train line: Repression and collusion along Europe's biggest infrastructure project. In: Dunlap, A., & Brock, A. (eds.) *Enforcing Ecocide Power, Policing & Planetary Militarization*. Basingstoke, Palgrave Macmillan.

Brock, G. 2018. Sufficiency and needs-based approaches. In: Olsaretti, S. (ed.) *The Oxford Handbook of Distributive Justice*. Oxford, Oxford University Press.

Broome, J. 2012. *Climate Matters: Ethics in a Warming World.* Cambridge, The White Horse Press.

Brown, D., Wheatley, H., Chaitanya, K., & Marshall, J. 2020. *A Green Economic Stimulus for Housing: The Macroeconomic Impacts of a UK Whole House Retrofit Programme.* London, New Economics Foundation.

Brundtland, G. H. 1987. *Our Common Future: Report of the World Commission on Environment and Development*. Geneva, UN-Dokument A/42/427.

Buck, H. J. 2021. *Ending Fossil Fuels: Why Net Zero Is Not Enough*. London, Verso Books.

Buller, A., & Braun, B. 2021. *Under New Management: Share Ownership and the Rise of Asset Manager Capitalism*. London, Common-Wealth.

Burke, E. 1951. *Reflections on the Revolution in France*. New York, E.P. Dutton.

Burke, M. J., & Stephens, J. C. 2018. Energy democracy: Goals and policy instruments for sociotechnical transitions. *Energy Research & Social Science*, 33, 35–48.

Burns, D. 2007. *Systemic Action Research: A Strategy for Whole System Change*. Bristol, The Policy Press.

Butler, J. 2020. *The Force of Non-Violence*. London, Verso.

Caddick, D., Tims, S., & Stirling, A. 2022. *23.4 Million People Unable to Afford the Cost of Living this Spring*. New Economics Foundation. [Online]. [Accessed 23 March 2022]. https://neweconomics.org/2022/03/23-4-million-people-unable-to-afford-the-cost-of-living

Cammaerts, B., Decillia, B., Viera Magalhães, J., & Jimenez-Martínez, C. 2016. *Journalistic Representations of Jeremy Corbyn in the British Press: From "Watchdog" to "Attackdog"*. London, London School of Economics.

Caney, S. 2014. Climate change, intergenerational equity and the social discount rate. *Politics, Philosophy & Economics*, 13(4), 320–342.

Caplin, A., & Leahy, J. 2004. The social discount rate. *Journal of Political Economy*, 112(6), 1257–1268.

Carrington, D. 2022. Car tyres produce vastly more particle pollution than exhausts, test show. *The Guardian*. [Online]. [Accessed September 2021]. https://www.theguardian.com/environment/2022/jun/03/car-tyres-produce-more-particle-pollution-than-exhausts-tests-show?CMP=Share_AndroidApp_Other

Centeno, M. A. 1993. The new Leviathan: The dynamics and limits of technocracy. *Theory and Society*, 22(3), 307–335.

Cerdas, F., Egede, P., & Herrmann, C. 2018. LCA of electromobility. In: Hauschild, M. Z. Rosenbaum, R. K., & Olsen, S. I. (eds.) *Life Cycle Assessment: Theory and Practice*. Berlin, Springer. [Online]. [Accessed December 2022]. https://www.youtube.com/watch?v=VlYEQ3JYMuc

Cesnulaityte, I., Davies, L., Sarabi, R., Camner, L., & González, T. 2020. *What If the People Governed? Lessons from Chile, Paris and California, 2022*. Hannah Arendt Center for Politics and Humanities at Bard College. [Online]. [Accessed December 2022]. https://www.youtube.com/watch?v=VlYEQ3JYMuc

CDUT (Canada-Déline Uranium Table). 2002. *Action Plan*. Ontario, CDUT.

CDUT (Canada-Déline Uranium Table). 2005. *Canada-Déline Uranium Table – Final Report: Concerning Health and Environmental Issues Related to the Port Radium Mine*. Ottawa, CDUT.

Chancel, L., & Piketty, T. 2015. *Carbon and Inequality: From Kyoto to Paris*. Paris, Paris School of Economics.

Char, N. L., & Csik, B. J. 1987. Nuclear power development: History and outlook. *IAEA Bulletin*, 3, 19–25.
Chilvers, J., & Burgess, J. 2008. Power relations: The politics of risk and procedure in nuclear waste governance. *Environment and Planning A*, 40, 1881–1900.
Christophers, B. 2020. *Rentier Capitalism: Who Owns the Economy, and Who Pays for It?* London, Verso.
Christophers, B. 2023. *Our Lives in Their Portfolios: Why Asset Managers Own the World*. London, Verso Books.
Churchman, C.W., 1967. Wicked problems in design Management Science. 14, B141–B142.
Cisco, G., & Gatto, A. 2021. Climate justice in an intergenerational sustainability framework: A stochastic OLG model. *Economies*, 9, 1–13.
Clayton, S., Devine-Wright, P., Stern, P. C., Whitmarsh, L., Carrico, A., Steg, L., Swim, J., & Bonnes, M. 2015. Psychological research and global climate change. *Nature Climate Change*, 5, 640–646.
CNSC (Canadian Nuclear Safety Commission). 2012. *Canada's Historical Role in Developing Nuclear Weapons*. [Online]. [Accessed June 2023]. http://www.nuclearsafety.gc.ca/eng/resources/fact-sheets/Canadas-contribution-to-nuclear-weapons-development.cfm
Collingridge, D. 1980. *The Social Control of Technology*. New York, St. Martin's Press.
CoRWM (Committee on Radioactive Waste Management). 2006. *Managing our Radioactive Waste Safely: CoRWM's Recommendations to Government*. [Online]. https://assets.publishing.service.gov.uk/government/uploads/system/uploads/attachment_data/file/294118/700__CoRWM_July_2006_Recommendations_to_Government_pdf.pdf
Cotton, M. 2013. Deliberating intergenerational environmental equity: A pragmatic, future studies approach. *Environmental Values*, 22, 317–337.
Cotton, M. 2014. *Ethics and Technology Assessment: A Participatory Approach*. Berlin, Springer.
Cotton, M. 2017. *Nuclear Waste Politics: An Incrementalist Perspective*. Abingdon, Routledge.
Cotton, M. 2018. Environmental justice as scalar parity: Lessons from nuclear waste management. *Social Justice Research*, 31, 238–259.
Cotton, M. 2020. Nuclear power and environmental justice – the case for political equality. In: Kaijser, A., Lehtonen, M., Meyer, J. H., & Rubio-Varas, M. D. M. (eds.) *Engaging the Atom: The History of Nuclear Energy and Society in Europe from the 1950s to the Present*. Morgantown, VA, West Virginia University Press.
Cotton, M., & Rowe, G. 2019. *Backcasting Futures for Nuclear Energy and Society: A Qualitative Analysis of European Stakeholder Perspectives*. Brussels, European Commission.
Crist, E. 2016. The reaches of freedom: A response to an ecomodernist manifesto. *Environmental Humanities*, 7(1), 245–254.
Cronon, W. 1996. The trouble with wilderness: Or, getting back to the wrong nature. *Environmental History*, 1(1), 7–28.
Curly, A. 2014. The origin of legibility: Rethinking colonialism and resistance among the Navajo people, 1868–1937. In: Lloyd, L. L. (ed.) *Diné Perspectives Revitalizing and Reclaiming Navajo Thought*. Tucson, University of Arizona Press.

Dafermos, Y., Gabor, D., & Michell, J. 2021. The Wall Street Consensus in pandemic times: What does it mean for climate-aligned development? *Canadian Journal of Development Studies/Revue canadienne d'études du développement*, 42, 238–251.

Daggett, C. N. 2019. *The Birth of Energy: Fossil Fuels, Thermodynamics and the Politics of Work*. Durham, NC, Duke University Press.

Dalquist, S. 2004. *Timeline: A Chronology of Public Opinion on Nuclear Power in the United States and United Kingdom*. Boston, MA, MIT. [Online]. [Accessed January 2024]. https://web.mit.edu/skd/www/dalquist-nuclear-opinion.pdf

Davidson, M. D. 2012. Distributive justice in the international regulation of global ecosystem services. *Global Environmental Change*, 22(4), 852–861.

Davies, K. J. 2007. Alive: Culture, Sustainability and Intergenerational Democracy. *University of Melbourne Refereed E-Journal*.

Davies, K. J. 2012. *Intergenerational Democracy: Rethinking Sustainable Development*. London, Common Ground Publishing.

Davies, K. J. 2015. Kastom, climate change and intergenerational democracy: Experiences from Vanuatu. In: Filho, W. L. (ed.) *Climate Change in the Asia-Pacific Region*. Singapore, Springer.

Davies, K. J., Tabucannon, G. M., & Box, P. 2016. Children, climate change, and the intergenerational right to a viable future. In: Ansell, N., & Klocker, N. (eds.) *Geographies of Global Issues: Change and Threat*. Berlin, Springer.

Davies, W., Dutta, S. J., Taylor, N., & Tazzoli, M. 2022. *Unprecedented: How Covid-19 Revealed the Politics of Our Economy*. London, Goldsmiths Press.

Dean, M., & Zamora, D. 2021. *The Last Man Takes LSD: Foucault and the End of Revolution* London, Verso.

Decanio, S. J. 2003. *Economic Models of Climate Change: A Critique*. Basingstoke, Palgrave Macmillan.

DECC (Department of Energy and Climate Change). 2011. *Waste Transfer Pricing Methodology for the Disposal of Higher Activity Waste from New Nuclear Power Stations*. [Online]. [Accessed August 2023]. https://assets.publishing.service.gov.uk/government/uploads/system/uploads/attachment_data/file/42629/3798-waste-transfer-pricing-methodology.pdf

DECC (Department of Energy and Climate Change). 2014. *Implementing a Geological Disposal Facility*. [Online]. [Accessed June 2023]. https://assets.publishing.service.gov.uk/media/5a7df4ca40f0b62305b7fdbb/GDF_White_Paper_FINAL.pdf

Deleuze, G. 1990. *Expressionism in Philosophy: Spinoza*. New York, Zone Press.

Deleuze, G., & Guattari, F. 1987. *A Thousand Plateaus: Capitalism and Schizophrenia*. Minneapolis and St Paul, University of Minnesota Press.

Deloitte. 2021. *Technology Futures Report 2021: Projecting the Possible. Navigating What's Next*. New York, Deloitte.

Descartes, R. 2017. *Meditations on First Philosophy in Which Are Demonstrated the Existence of God and the Distinction Between the Human Soul and Body*. [Online]. [Accessed March 2024]. https://www.earlymoderntexts.com/assets/pdfs/descartes1641.pdf

Desnez. 2023, Winter. *DESNZ Public Attitudes Tracker: Energy Infrastructure and Energy Sources*. London, Department for Energy Security and Net-Zero.
Dorling, D. 2020. Can economic inequality be reduced? Challenges and signs of hope in 2019. In: Galvin, R. (ed.) *Inequality and Energy: How Extremes of Wealth and Poverty in High Income Countries Affect CO_2 Emissions and Access to Energy*. London, Academic Press.
Dorling, D. 2023. *Shattered Nation: Inequality and the Geography of a Failing State*. London, Verso Books.
Dryzek, J. S. 2000. *Deliberative Democracy and Beyond: Liberals, Critics, Contestations*. New York, Oxford University Press.
Dryzek, J. S. 2013. *The Politics of the Earth*. New York, Oxford University Press.
Dryzek, J. S. 2014. Institutions for the Anthropocene: Governance in a changing Earth system. *British Journal of Political Science*, 46, 937–956.
Dryzek, J. S., Goodin, R. E., Tucker, A., & Reber, B. 2008. Promethean elites encounter precautionary publics: The case of GM foods. *Science, Technology & Human Values*, 34, 263–288.
Dryzek, J. S., & Pickering, J. 2019. *The Politics of the Anthropocene*. Oxford, Oxford University Press.
Dryzek, J. S., & Stephenson, H. 2011. Global democracy and earth system governance. *Ecological Economics*, 70, 1865–1874.
DTI (Department for Trade and Industry). 2003. *Our Energy Future: Creating a Low-Carbon Economy*. CM5761. [Online]. [Accessed June 2023]. https://assets.publishing.service.gov.uk/government/uploads/system/uploads/attachment_data/file/272061/5761.pdf.
Dunbar-Ortiz, R. 2023. *An Indigenous Peoples' History of the United States*. Boston, Beacon Press.
Dunlap, A., & Arce, M. C. 2021. 'Murderous energy' in Oaxaca, Mexico: Wind factories, territorial struggle and social warfare. *The Journal of Peasant Studies*, 1–26.
Edwards, T. S., & Kelton, P. 2020. Germs, genocides, and America's indigenous peoples. *Journal of American History*, 107(1), 52–76.
Endres, D. 2009. The rhetoric of nuclear colonialism: Rhetorical exclusion of American Indian arguments in the Yucca Mountain nuclear waste siting decision. *Communication and Critical/Cultural Studies*, 6(1), 39–60.
Engle, J., Agyeman, J., & Chung-Tiam-Fook, T. 2022. Imagine shaping cities as if people, land, and nature were sacred. In: Engle, J., Agyeman, J., & Chung-Tiam-Fook, T. (eds.) *Sacred Civics: Building Seven Generation Cities*. London, Earthscan.
Ercan, S. A., & Dryzek, J. S. 2015. The reach of deliberative democracy. *Policy Studies*, 36(3), 241–248.
Erickson, M. 2016. *Science, Culture and Society: Understanding Science in the 21st Century*. London, John Wiley & Sons.
Escobar, A. 2010. Worlds and knowledges otherwise: The Latin American modernity/coloniality research program. In: Mignolo, W. D., & Escobar, A. (eds.) *Globalization and the Decolonial Option*. Abingdon, Routledge.

Estes, N. 2019. *Our History Is the Future: Standing Rock Versus the Dakota Access Pipeline, and the Long Tradition of Indigenous Resistance*. London, Verso.

Eversberg, D., & Schmelzer, M. 2023, May 3. Degrowth and masculinities: Towards a gendered understanding of degrowth subjectivities. *Degrowth Journal*, 1, 00009.

Fairlie, I. 2014. A hypothesis to explain childhood cancers near nuclear power plants. *Journal of Environmental Radioactivity*, 133, 10–17.

Falk, A., & Fischbacher, U. 2006. A theory of reciprocity. *Games and Economic Behavior*, 54(2), 293–315.

Feyerabend, P. 1977. Rationalism, relativism and scientific method. *Philosophy in Context*, 6, 7–19.

Fichtner, J., Heemskerk, E. M., & Garcia-Bernardo, J. 2017. Hidden power of the big three? Passive index funds, re-concentration of corporate ownership, and new financial risk. *Business and Politics*, 19(2), 298–326.

Fischer, F. 2000. *Citizens, Experts, and the Environment: The Politics of Local Knowledge*. Durham, NC, Duke University Press.

Fisher, M. 2009. *Capitalist Realism: Is There No Alternative?* London, Zed Books.

Flynn, J. 2003. Nuclear stigma. In: Pidgeon, N., Kasperson, R. E., & Slovic, P. (eds.) *The Social Amplification of Risk*. Cambridge, Cambridge University Press.

Foucault, M. 1994. What is an author? In: Faubion, J. D. (ed.) *Aesthetics, Method, and Epistemology: Essential Works of Michel Foucault 1954–1984 Volume 2*. New York, The New York Press.

Foucault, M. 1996. What is enlightenment? In: Rabinow, P. (ed.) *Ethics: Subjectivity and Truth: Essential Works of Michel Foucault 1954–1984 Volume 1*. New York, The New York Press.

Frank, J. 2022. *Atomic Days: The Untold Story of the Most Toxic Place in America*. Chicago, Haymarket Books.

Frank, S. 2016. The "force in the thing": Mauss' nonauthoritarian sociality in the gift. *HAU: Journal of Ethnographic Theory*, 6(2), 255–277.

Frankfurt, H. 1987. Equality as a moral ideal. *Ethics*, 98(1), 21–43.

Fraser, N. 2014. Behind Marx's hidden abode: For an expanded conception of capitalism. *New Left Review*, 86, 55–72.

Fraser, N. 2021. Climates of capital. *New Left Review*, 127, 94–127.

Fritsch, M. 2018. *Taking Turns with the Earth: Phenomenology, Deconstruction, and Intergenerational Justice*. Stanford, CA, Stanford University Press.

Funtowicz, S. O., & Ravetz, J. R. 1993. Science for the post-normal age. *Futures*, 25, 739–755.

Gabor, A. 2013. The savings principle and inter-generational justice. *European Journal of Science and Theology*, 9(2), 299–308.

Gabor, D. 2022. *Wall Street Consensus at COP27: The Derisking Roll-Out at COP27*. Phenomenal World. [Online]. [Accessed November 2022]. https://www.phenomenalworld.org/analysis/the-wall-street-consensus-at-cop27/

Galvin, R. 2020. Economic inequality, energy justice and the meaning of life. In: Galvin, R. (ed.) *Inequality and Energy: How Extremes of Wealth and Poverty in High Income Countries Affect CO_2 Emissions and Access to Energy*. Cambridge, MA, Academic Press.

Geels, F. 2014. Regime resistance against low-carbon transitions: Introducing politics and power into the multi-level perspective. *Theory, Culture & Society*, 31, 21–40.

Gelderloos, G. 2022. *The Solutions Are Already Here: Tactics for Ecological Revolutions from Below.* London, Pluto Press.

Genus, A., & Stirling, A. 2018. Collingridge and the dilemma of control: Towards responsible and accountable innovation. *Research Policy*, 47(1), 61–69.

Gilbert, S. F., 2020. Developmental symbiosis facilitates the multiple origins of herbivory. *Evolution & Development*, 22(1–2), 154–164.

Gilmore, R. W. 2023. *Abolition Geography: Essays Toward Liberation.* London, Verso.

Ginn, F., Bastian, M., Farrier, D., & Kidwell, J. 2018. Introduction: Unexpected encounters with deep time. *Environmental Humanities*, 10, 213–225.

Godelier, M. 1999. *The Enigma of the Gift*. Cambridge, Polity Press.

Gómez, M. 2022. *Nuclear Nuevo México: Colonialism and the Effects of the Nuclear Industrial Complex on Nuevomexicanos*. Tucson, University of Arizona Press.

Goodfellow, M. 2019. *Hostile Environment: How Immigrants Became Scapegoats*. London, Verso.

GOTEC (Guardians of the East Coast). 2024a. *The People Are Not Well Enough Informed to Vote on GDF*. [Online]. [Accessed March 2024]. https://www.gotec.org.uk/post/the-people-are-not-well-enough-informed-to-vote-on-gdf

GOTEC (Guardians of the East Coast). 2024b. *A Tale of Two Search Areas*. [Online]. [Accessed March 2024]. https://www.gotec.org.uk/post/a-tale-of-two-search-areas

GOTEC (Guardians of the East Coast). 2024c. *Test of Public Support*. [Online]. [Accessed March 2024]. https://www.gotec.org.uk/general-5-1

GOV.UK. 2024a. *About Us: Nuclear Waste Services*. [Online]. [Accessed June 2023]. https://www.gov.uk/government/organisations/nuclear-waste-services/about

GOV.UK. 2024b. *Biggest Expansion of Nuclear Power for 70 Years to Create Jobs, Reduce Bills and Strengthen Britain's Energy Security*. [Press release]. [Accessed March 2024]. https://www.gov.uk/government/news/biggest-expansion-of-nuclear-power-for-70-years-to-create-jobs-reduce-bills-and-strengthen-britains-energy-security

Gowing, M. 1974. *Independence and Deterrence: Britain and Atomic Energy, 1945–1952*. Basingstoke, Palgrave Macmillan.

Graeber, D. 2001. *Toward and Anthropological Theory of Value: The False Coin of Our Own Dreams*. New York, Palgrave MacMillan.

Graeber, D. 2007. There never was a west: Or, democracy emerges from the spaces in between. In: Graeber, D. (ed.) *Possibilities: Essays on Hierarchy, Rebellion and Desire*. Stirling, CA, AK Press.

Graeber, D. 2011. *Debt: The First Five Thousand Years*. London, Melville House.
Graeber, D. 2018. *Bullshit Jobs: A Theory*. London, Simon & Schuster.
Graeber, D., & Wengrow, D. 2021. *The Dawn of Everything: A New History of Humanity*. London, Penguin.
Gray, J., Jones, S. R., & Smith, A. D. 1995. Discharges to the environment from the Sellafield site, 1951–1992. *Journal of Radiological Protection*, 15(2), 99.
Green, D., & Raygorodetsky, G. 2010. Indigenous knowledge of a changing climate. *Climatic Change*, 100(2), 239.
Greene, D. 2022. Landlords of the Internet: Big data and big real estate. *Social Studies of Science*, 1, 904–927.
Grönlund, K., Setälä, M., & Herne, K. 2010. Deliberation and civic virtue: Lessons from a citizen deliberation experiment. *European Political Science Review*, 2, 95–117.
Groves, C. 2009. Future ethics: Risk, care and non-reciprocal responsibility. *Journal of Global Ethics*, 5(1), 17–31.
Gumbrecht, H. U. 2014. *Our Broad Present: Time and Contemporary Culture*. Stamford, CA, Columbia University Press.
Habermas, J. 1972. *Knowledge and Human Interests*. London, Heinemann Educational.
Habermas, J. 1996. *Between Facts and Norms: Discourse Theory of Law and Democracy*. Cambridge, MA, MIT Press.
Hadjilambrinos, C. 1990. Nuclear power as an ethical issue: Utilitarian ethics and Egalitarian responses. *Bulletin of Science, Technology & Society*, 10, 282–289.
Hajer, M. A. 1995. *The Politics of Environmental Discourse: Ecological Modernization and the Policy Process*. Oxford, Clarendon Press.
Hamilton, P. F. 2004. *Pandora's Star*. Basingstoke, Macmillan Publishers.
Hamilton, P. F. 2005. *Judas Unchained*. Basingstoke, Macmillan Publishers.
Hansen, J. 2009. *Storms of My Grandchildren: The Truth about the Coming Climate Catastrophe and Our Last Chance to Save Humanity*. London, Bloomsbury.
Haraway, D. J. 1988. Situated knowledges: The science question in feminism and the privilege of partial perspective. *Feminist Studies*, 14, 575–599.
Haraway, D. J. 2016. *Staying with the Trouble: Making Kin in the Chthulucene*. Durham, NC, Duke University Press.
Hardin, G. 1968. The tragedy of the commons: The population problem has no technical solution; it requires a fundamental extension in morality. *Science*, 162(3859), 1243–1248.
Harris, M. 2023. *Palo Alto: A History of California, Capitalism, and the World*. New York, Boston, and London, Little, Brown and Company.
Hart, D. W., & Cornia, G. C. 1994. From intergenerational equity to intergenerational stewardship. *International Journal of Public Administration*, 17(12), 2367–2389.
Harvey, D. 2007. *A Brief History of Neoliberalism*. Oxford, Oxford University Press.
Haszeldine, S., & Smythe, D. 1997. Why was Sellafield rejected as a disposal site for radioactive waste? *Geoscientist*, 7(7), 18–20.

Hayes, S., & O'Neill, S. 2021. The Greta effect: Visualising climate protest in UK media and the Getty images collections. *Global Environmental Change*, 71, 102392.

Haywood, L., Leroutier, M., & Pietzcker, R. 2023. Why investing in new nuclear plants is bad for the climate. *Joule*, 7(8), 1675–1678.

Head, B. W. 2022. *Wicked Problems in Public Policy: Understanding and Responding to Complex Challenges*. Berlin, Springer Nature.

Heath, J. S. 2013. The nuclear rabbit hole: Environmental justice for women, children, and the unborn. *The McNair Scholars Journal of the University of Washington*, 13, 95–112.

Hecht, G. 2012. *Being Nuclear: Africans and the Global Uranium Trade*. Cambridge, MA, MIT Press.

Heffron, R. J., Mccauley, D., & Sovacool, B. K. 2015. Resolving society's energy trilemma through the energy justice metric. *Energy Policy*, 87, 168–176.

Henningson, D. 2006. *Somba Ke – The Money Place*. [Online]. [Accessed December 2022]. https://www.youtube.com/watch?v=hzhQBNmIXzA

Herman, E. S., & Chomsky, N. 2002. *Manufacturing Consent: The Political Economy of the Mass Media*. New York, Pantheon Books.

Hickel, J. 2018. The Nobel prize for climate catastrophe. *Foreign Policy*. [Online]. [Accessed November 2022]. https://foreignpolicy.com/2018/12/06/the-nobel-prize-for-climate-catastrophe/

Hickel, J. 2020. *Less Is More: How Degrowth Will Save the World*, London, Random House.

Hickel, J., Dorninger, C., Wieland, H., & Suwandi, I. 2022. Imperialist appropriation in the world economy: Drain from the global South through unequal exchange, 1990–2015. *Global Environmental Change*, 73, 102467.

Hickel, J., & Kallis, G. 2019. Is green growth possible? *New Political Economy*, 24, 469–486.

Hickel, J., Sullivan, D., & Zoomkawala, H. 2021. Plunder in the post-colonial era: Quantifying drain from the Global South through unequal exchange, 1960–2018. *New Political Economy*, 26(6), 1030–1047.

Hiskes, R. P. 2005. The right to a green future: Human rights, environmentalism, and intergenerational justice. *Human Rights Quarterly*, 27, 1346–1364.

HM (Her Majesty's) Government. 2008. *Meeting the Energy Challenge*. [Online]. [Accessed November 2022]. https://assets.publishing.service.gov.uk/media/5a7490ace5274a44083b7b15/7296.pdf

Hobbes, T. 1914. *Leviathan*. London, J.M. Dent & Sons.

Homans, G. C. 1974. *Social Behavior: Its Elementary Forms* (Revised ed.). London, Harcourt Brace.

Hookway, B. R. 1984. Radioactive waste management: 1963–1984. *Journal of the Society for Radiological Protection*, 4(3), 122.

Horn-Miller, K. 2013. What does Indigenous participatory democracy look like? Kahnawa:ke's community decision making process. *Review of Constitutional Studies*, 18, 111–132.

Hornborg, A. 2009. Zero-sum world: Challenges in conceptualizing environmental load displacement and ecologically unequal exchange in the world-system. *International Journal of Comparative Sociology*, 50(3–4), 237–262.

Howard, S. 2019. Canada's uranium highway: Victims and perpetrators. *Cape Breton Spectator*. [Online]. [Accessed March 2024]. https://capebreton-spectator.com/2019/08/07/canadas-uranium-dene-bomb/
Howarth, R. B. 1998. An overlapping generations model of climate-economy interactions. *Scandinavian Journal of Economics*, 100, 575–591.
Huber, M. T. 2013. *Lifeblood: Oil, Freedom, and the Forces of Capital*. Minneapolis and St Paul, University of Minnesota Press.
Hudson, K. 2024. *MPs Must Oppose US–UK Nuclear Arms Accord*. Declassified UK. [Online]. [Accessed March 2024]. https://www.declassifieduk.org/mps-must-oppose-us-uk-nuclear-arms-accord/
Hunhold, C. 2002. Canada's low-level radioactive waste disposal problem: Voluntarism reconsidered. *Environmental Politics*, 11, 49–72.
Hustak, C., & Myers, N. 2012. Involutionary momentum: Affective ecologies and the sciences of plant/insect encounters. *Differences*, 23(3), 74–118.
IAEA (International Atomic Energy Agency). 1999. *Inventory of Radioactive Material Entering the Marine Environment: Sea Disposal of Radioactive Waste*. IAEA. [Online]. [Accessed March 2024]. https://www-pub.iaea.org/MTCD/Publications/PDF/te_588_web.pdf
Ialenti, V. 2020a. Spectres of Seppo: The afterlives of Finland's nuclear waste experts. *Journal of the Royal Anthropological Institute*, 26, 251–268.
Ialenti, V. 2020b. *Deep Time Reckoning: How Future Thinking Can Help Earth Now*. Cambridge, MA, The MIT Press.
Inglehart, R., Foa, R., Peterson, C., & Welzel, C. 2008. Development, freedom, and rising happiness: A global perspective (1981–2007). *Perspectives on Psychological Science*, 3(4), 264–285.
IPBES (Intergovernmental Science-Policy Platform on Biodiversity and Ecosystem Services). 2019. *Nature's Dangerous Decline 'Unprecedented' Species Extinction Rates 'Accelerating'*. [Press release]. [Accessed May 2019]. https://www.ipbes.net/
IPCC. 2022. Climate change 2022: Impacts, adaptation, and vulnerability. In *Contribution of Working Group II to the Sixth Assessment Report of the Intergovernmental Panel on Climate Change*. Cambridge, Cambridge University Press.
Jackson, I. 2011. *Research Report: Subsidy Assessment of Waste Transfer Pricing for Disposal of Spent Fuel from New Nuclear Power Stations*. Jackson Consulting. [Online]. [Accessed May 2023]. https://storage.googleapis.com/gpuk-old-wp-site/wp-content/uploads/2017/07/FUP-Subsidy-Report-Mar2011.pdf
Jackson, T. 2017. *Prosperity Without Growth: Economics for a Finite Planet*. Abingdon, Routledge.
Jacobsson, D. 2021. Young vs old? Truancy or new radical politics? Journalistic discourses about social protests in relation to the climate crisis. *Critical Discourse Studies*, 18, 481–497.
Jasanoff, S. 1999. The songlines of risk. *Environmental Values*, 8, 135–152.
Jenkins, K. Mccauley, D. Heffron, R. Stephan, H., & Rehner, R. 2016. Energy justice: A conceptual review. *Energy Research & Social Science*, 11, 174–182.
Johnstone, P., & Stirling, A. 2021, Summer. Hidden military implications of 'building back' with new nuclear in the UK. *Responsible Science*, 3.

[Online]. [Accessed March 2024]. https://www.sgr.org.uk/sites/default/files/2021-09/SGR_RS03_2021_Johnstone%2BStirling.pdf

Jones, N., O'Brien, M., & Ryan, T. 2018. Representation of future generations in United Kingdom policy-making. *Futures*, 102, 153–163.

Kagel, J. H. 2018. Cooperation through communication: Teams and individuals in finitely repeated Prisoners' dilemma games. *Journal of Economic Behavior & Organization*, 146.

Kari, M., Kojo, M., & Litmanen, T. 2010. *Community Divided: Adaptation and Aversion towards the Spent Nuclear Fuel Repository in Eurajoki and Its Neighbouring Municipalities.* Tampere, University of Jyvaskyla/University of Tampere.

Kasperson, R. E., Renn, O., Slovic, P., Brown, H. S., Emel, J., Goble, R., Kasperson, J. X., & Ratick, S. 1988. The social amplification of risk: A conceptual framework. *Risk Analysis*, 8, 177–187.

Keen, S. 2020. The appallingly bad neoclassical economics of climate change. *Globalizations*. doi: 10.1080/14747731.2020.1807856.

Kenner, D. 2019. *Carbon Inequality: The Role of the Richest in Climate Change*. Abingdon, Routledge.

Kessler, O. 2015. Ignorance and the sociology of economics. In: Gross, M., & Mcgoey, L. (eds.) *Routledge International Handbook of Ignorance Studies* (pp. 338–348). Abingdon, Routledge.

Keynes, J. M. 1930. Economic possibilities for our grandchildren. In: *Essays in Persuasion* (pp. 321–332). London, Palgrave Macmillan UK.

Kimmerer, R. 2013. *Braiding Sweetgrass: Indigenous Wisdom, Scientific Knowledge and the Teachings of Plants.* London, Milkweeds Editions.

Knappe, H., & Renn, O. 2022. Politicization of intergenerational justice: How youth actors translate sustainable futures. *European Journal of Futures Research*, 10, 1–11.

Koirala, P., Timilsina, R. R., & Kotani, K. 2021. Deliberative forms of democracy and intergenerational sustainability dilemma. *Sustainability*, 13, 7377.

Kojo, M. 2009. The strategy of site selection for the spent nuclear fuel repository in Finland. In: Kojo, M., & Litmanen, T. (eds.) *The Renewal of Nuclear Power in Finland.* Basingstoke, Palgrave Macmillan.

Kojo, M., Kari, M., & Litmanen, T. 2010. The socio-economic and communication challenges of spent nuclear fuel management in Finland: The post site selection phase of the repository project in Eurajoki. *Progress in Nuclear Energy*, 52, 168–176.

Kojo, M., Kari, M., Litmanen, T., Vilhunen, T., & Lehtonen, M. 2020. The critical Swedes and the consensual Finns: Leading newspapers as watchdogs or lapdogs of nuclear waste repository licensing? *Energy Research & Social Science*, 61, 101354.

Koplow, D. 2011. *Nuclear Power: Still Not Viable Without Subsidies*, Union of Concerned Scientists. Earth Track. [Online]. [Accessed November 2023]. https://www.ucsusa.org/sites/default/files/2019-09/nuclear_subsidies_report.pdf

Koram, K. 2022. *Uncommon Wealth: Britain and the Aftermath of Empire*. London, John Murray.

Krall, M. L., Macfarlane, A. M., & Ewing, R. C. 2021. Nuclear waste from small modular reactors. *Proceedings of the National Academy of Sciences*, 119, e2111833119.

Krishnan, A. 2022. Unsettling the coloniality of foresight. In: Engle, J., Agyeman, J., & Chung-Tiam-Fook, T. (eds.) *Sacred Civics: Building Seven Generation Cities*. London, Earthscan.

Kristiansen, S. 2017. Characteristics of the mass media's coverage of nuclear energy and its risk: A literature review. *Sociology Compass*, 11, e12490.

Kuhn, R. G. 1998. Social and political issues in siting a nuclear-fuel waste disposal facility in Ontario, Canada. *The Canadian Geographer*, 42, 14–28.

Lake, M., & Reynolds, H. 2008. *Drawing the Global Colour Line: White Men's Countries and the Question of Racial Equality*. Melbourne, Melbourne University Publishing.

Lawrence, P. 2021. Justifying institutions for future generations based on the mitigation of bias and intergenerational justice. In: Lawrence, P., & Linehan, J. (eds.) *Giving Future Generations a Voice Normative Frameworks, Institutions and Practice*. Cheltenham, Edward Elgar Publishing.

Lawson, A., & Isaac, A. 2023. 'Dirty 30' and its toxic siblings: The most dangerous parts of the Sellafield nuclear site. *The Guardian*. [Online]. [Accessed November 2023]. https://www.theguardian.com/business/2023/dec/05/dirty-30-dangerous-sellafield-nuclear-site-ponds-safety-fears

Lazard. 2020. *Lazard's Levelized Cost of Energy Analysis – Version 14*.0. London, Lazard.

Lehtonen, M. 2022. Building promises of small modular reactors one conference at a time. *Bulletin of the Atomic Scientists*. [Online]. [Accessed November2023].https://thebulletin.org/2022/12/building-promises-of-small-modular-reactors-one-conference-at-a-time/

Lehtonen, M., & Kojo, M. 2019. The role and functions of community benefit schemes: A comparison of the Finnish and French nuclear waste disposal projects. In: Brunnengräber, A., & Di Nucci, M. R. (eds.) *Conflicts, Participation and Acceptability in Nuclear Waste Governance: An International Comparison Volume III*. Heidelberg, Springer VS.

Lehtonen, M., Kojo, M., Kari, M., & Litmanen, T. 2021. Healthy mistrust or complacent confidence? Civic vigilance in the reporting by leading newspapers on nuclear waste disposal in Finland and France. *Risks Hazards Crisis Public Policy*, 12, 130–157.

Lemos, M. C., & Agrawal, A. 2006. Environmental governance. *Annual Review of Environmental Resources*, 31, 297–325.

Lennon, B., Dunphy, N. P., & Sanvicente, E. 2019. Community acceptability and the energy transition: A citizens' perspective. *Energy, Sustainability and Society*, 9, 1–18.

Lenton, T. M., Held, H., Kriegler, E., Hall, J. W., Lucht, W., Rahmstorf, S., & Schellnhuber, H. J. 2008. Supplement to tipping elements in the earth's climate system. *Proceedings of the National Academy of Sciences*, 105(6), 1786–1793.

Leonard, T. C. 2009. American economic reform in the progressive era: Its foundational beliefs and their relation to eugenics. *History of Political Economy*, 41(1), 109–141.

Lewis, S. 2022. *Abolish the Family: A Manifesto for Care and Liberation*. London, Verso Books.

Lewis, S. L., & Maslin, M. A. 2015. Defining the Anthropocene. *Nature*, 519, 171–180.

Locke, J. 2015. *The Second Treatise of Civil Government*. Peterborough, ON, Broadview Press.

Lockie, S. 2018. Privilege and responsibility in environmental justice research. *Environmental Sociology*, 4(2), 175–180.

Lohmann, L. 2013. Interpretative openness and climate action in an age of market environmentalism. In: Methman, C., Rothe, D., & Stephan, B. (eds.) *Deconstructing the Greenhouse: Interpretive Approaches to Global Climate Governance*. Abingdon, Routledge.

Lowry, R., & Peterson, M. 2011. Pure time preference. *Pacific Philosophical Quarterly*, 92(4), 490–508.

Lucas, K., Stokes, G., Bastiaanssen, J., & Burkinshaw, J. 2019. *Inequalities in Mobility and Access in the UK Transport System. Future of Mobility: Evidence Review*. London, Foresight, Government Office for Science.

Mackensie, D. 2006. *An Engine, Not a Camera: How Financial Models Shape Markets*. Cambridge, MA, MIT Press.

Mackerron, G. 2015. Multiple challenges: Nuclear waste governance in the United Kingdom. In: Brunnengräber, A., Di Nucci, M. R., Losada, A. M. I., Mez, L., & Schreurs, M. A. (eds.) *Nuclear Waste Governance: An International Comparison* (pp. 101–116). Fachmedien Wiesbaden, Springer.

Malm, A. 2016a. *Fossil Capital: The Rise of Steam Power and the Roots of Global Warming*. London, Verso.

Malm, A. 2016b. Who lit this fire? Approaching the history of the fossil economy. *Critical Historical Studies*, 3(2), 215–248.

Malm, A. 2021. *How to Blow Up a Pipeline*. London, Verso.

Malm, A., & The Zetkin Collective. 2021. *White Skin, Black Fuel: On the Danger of Fossil Fascism*. London, Verso.

Mann, B. A. 2001. *Native American Speakers of the Eastern Woodlands: Selected Speeches and Critical Analysis*. London, Greenwood Publishing Group.

Mann, B. A. 2017. Foreword: "And then they build monuments to you". In: Churchill, W. (ed.) *Welding Words Like Weapons: Selected Essays in Indigenism, 1995–2005*. Oakland, CA, PM Press.

Marmot, M., Allen, J., Boyce, T., Goldblatt, P., & Morrison, J. 2020. *Health Equity in England: The Marmot Review 10 Years On*. London, Institute of Health Equity.

Marya, R., & Patel, R. 2021. *Inflamed: Deep Medicine and the Anatomy of Injustice*. London, Penguin.

Marx, K. 1852. *The Eighteenth Brumaire of Louis Bonaparte*. [Online]. [Accessed November 2023]. https://www.marxists.org/archive/marx/works/1852/18th-brumaire/

Marx, P. 2022. *Road to Nowhere: What Silicon Valley Gets Wrong about the Future of Transportation*. London, Verso Books.

Mather, J. D., Gray, D. A., Greenwood, P. B. 1979. Burying Britains's radioactive waste: The geological areas under investigation. *Nature*, 281(5730), 332–334.

Mau, S. 2023. *Mute Compulsion: A Marxist Theory of the Economic Power of Capital*. London, Verso Books.
Mauss, M. 2002. *The Gift: The Form and Reason for Exchange in Archaic Societies*. Abingdon, Routledge.
Mbembé, A. 2019. *Necropolitics*. Durham, NC, Duke University Press.
Mcarthur, D., & Reeves, A. 2019. The rhetoric of recessions: How British newspapers talk about the poor when unemployment rises. *Sociology*, 53, 1005–1025.
Mccauley, D. A. Heffron, R. J. Stephan, H., & Jenkins, K. 2013. Advancing energy justice: The triumvirate of tenets. *International Energy Law Review*, 32, 107–110.
Mccormick, H. 2009. Intergenerational justice and the non-reciprocity problem. *Political Studies*, 57(2), 451–458.
Mcgoey, L. 2019. *The Unknowers: How Strategic Ignorance Rules the World*. London, Bloomsbury Publishing.
Mcquaig, L., & Brooks, N. 2013. *The Trouble with Billionaires: How the Super-Rich Hijacked the World (and How We Can Take It Back)*. London, Simon & Schuster.
Mendelsohn, R., Morisson, W., Schlesinger, M. E., & Andronova, N. G. 2000. Country-specific market impacts of climate change. *Climatic Change*, 45(3), 553–569. doi: 10.1023/A:1005598717174.
Merchant, C. 1989. *The Death of Nature: Women, Ecology, and the Scientific Revolution*. London, Harper One.
Methmann, C., Rothe, D., & Stephan, B. 2013. *Deconstructing the Greenhouse: Interpretive Approaches to Global Climate Governance*. Abingdon, Routledge.
Meyer, L. H. ed. 2017. *Intergenerational Justice*. Abingdon, Routledge.
Mies, M., & Shiva, V. 2014. *Ecofeminism*. London, Bloomsbury Publishing.
Mignolo, W. D. 2007. Delinking: The rhetoric of modernity, the logic of coloniality and the grammar of de-coloniality. *Cultural Studies*, 21, 449–514.
Milfont, T. L., Bain, P. G., Kashima, Y., Corral-Verdugo, V., Pasquali, C., Johansson, L. O., Guan, Y., Gouveia, V. V., Garðarsdóttir, R. B., Doron, G., & Bilewicz, M. 2018. On the relation between social dominance orientation and environmentalism: A 25-nation study. *Social Psychological and Personality Science*, 9, 802–814.
Mirowski, P. 2013. *Never Let a Serious Crisis Go to Waste: How Neoliberalism Survived the Financial Meltdown*. London, Verso Books.
Mirowski, P. 2020. The neoliberal Ersatz Nobel prize. In: Plehwe, D., Slobodian, Q., & Mirowski, P. (eds.) *Nine Lives of Neoliberalism* (pp. 219–254). London, Verso.
Mitchell, G. 2020. Did Truman ever regret his Hiroshima decision? | opinion. *Newsweek*.
Mitchell, T. 2013. *Carbon Democracy: Political Power in the Age of Oil*. London, Verso.
Möckel, B. 2023. "What has posterity ever done for me?": Future generations, intergenerational justice, and the chronopolitics of distant futures. *History and Theory*, 62(4), 66–68.
Moore, J. W. 2015. *Capitalism in the Web of Life: Ecology and the Accumulation of Capital*. London, Verso.

Mueller, G. 2021. *Breaking Things at Work: The Luddites Are Right about Why You Hate Your Job*. London, Verso.
Murdock, E. G. 2020. A history of environmental justice: Foundations, narratives, and perspectives. In: Coolsaet, B. (ed.) *Environmental Justice* (pp. 6–17). Abingdon, Routledge.
Nader, L. 2004. Social thought & commentary: The Harder path: Shifting gears. *Anthropological Quarterly*, 77, 771–791.
NDA (Nuclear Decommissioning Authority). 2023. *Annual Report and Accounts 2022/23*. [Online]. [Accessed February 2024]. https://assets.publishing.service.gov.uk/media/65097c6022a783001343e867/NDA_ARAC_2022_FINAL_web.pdf
Nevin, G., & Piccin, J. 2007. *Project to Explore Two-Way Communication with Aboriginal Peoples*. Ottawa, NWMO.
Newell, P., & Mulvaney, D. 2013. The political economy of the 'just transition'. *The Geographical Journal*, 179(2), 132–140.
Nielsen, L., & Axelsen, D. V. 2017. Capabilitarian sufficiency: Capabilities and social justice. *Journal of Human Development and Capabilities*, 18(1), 46–59.
Nirex. 2005. *Historic List of Possible Locations for a Radioactive Waste Repository*. Harwell, Nirex.
Nolden, C., Towers, L., Schmidt Rossade, D., Thomas, P., Speciale, G., & Watson, R. 2022. Tensions between decarbonisation and energy poverty priorities in liberalised electricity markets – evidence from community energy poverty alleviation in Great Britain. *Local Environment*, 27(9).
Nordhaus, W. D. 1991. To slow or not to slow: The economics of the greenhouse effect. *The Economic Journal*, 101(407), 920–937.
Nordhaus, W. D. 1994. Expert opinion on climatic change. *American Scientist*, 82(1), 45–51.
Nordhaus, W. D. 2006. Geography and macroeconomics: New data and new findings. *Proceedings of the National Academy of Sciences of the United States of America*, 103(10), 3510–3517.
Nordhaus, W. D. 2007. A review of the stern review on the economics of climate change. *Journal of Economic Literature*, 45, 686–702.
Nordhaus, W. D. 2018a. Nobel lecture in economic sciences. In: *Climate Change: The Ultimate Challenge for Economics*. [Online]. [Accessed February 2024]. https://www.nobelprize.org/uploads/2018/10/nordhaus-slides.pdf
Nordhaus, W. D. 2018b. Projections and uncertainties about climate change in an era of minimal climate policies. *American Economic Journal: Economic Policy*, 10(3), 333–360.
Nordhaus, W. D., & Boyer, J. 2000. *Warming the World: Economic Models of Global Warming*. Cambridge, MA, MIT Press.
Norgate, T., Haque, N., & Koltun, P. 2014. The impact of uranium ore grade on the greenhouse gas footprint of nuclear power. *Journal of Cleaner Production*, 84, 360–367.
Norton-Taylor, R. 2024. *Who Should We Bomb Next?* Declassified UK. [Online]. [Accessed February 2024]. https://www.declassifieduk.org/who-should-we-bomb-next/

Nussbaum, M. C. 2004. Capabilities as fundamental entitlements: Sen and social justice. *Feminist Economics*, 9, 33–59.
NWMO (Nuclear Waste Management Organisation). 2020. *Nuclear Waste Management Organization Reconciliation Baseline Evaluation Summary Report*. Ottawa, NWMO.
NWMO (Nuclear Waste Management Organisation). 2021a. *Guided by Science. Grounded in Knowledge. Committed to Partnership*. Annual Report 2021. Ottawa, NWMO.
NWMO (Nuclear Waste Management Organisation). 2021b. *Engaging Youth: A Report on NWMO Youth Engagement Activities*. Ottawa, NWMO.
NWMO (Nuclear Waste Management Organisation). 2022. *Implementing Adaptive Phased Management 2022 to 2026*. Ottawa, NWMO.
NWMO (Nuclear Waste Management Organisation). 2023a. *Ethical and Social Framework*. [Online]. [Accessed June 2023]. https://www.nwmo.sca/en/ABOUT-US/Who-We-Are/Our-Commitment/Ethical-and-Social-Framework
NWMO (Nuclear Waste Management Organisation). 2023b. *Who We Are*. [Online]. [Accessed June 2023]. https://www.nwmo.ca/en/ABOUT-US/Who-We-Are
NWMO (Nuclear Waste Management Organisation). 2023c. *How We're Governed*. Ottawa, NWMO.
NWMO (Nuclear Waste Management Organisation). 2023d. *About the Process*. Ottawa, NWMO.
NWS (Nuclear Waste Services). n.d. a. *Securing a Safer Future for Us All: Introducing Nuclear Waste Services*. Calderbridge Cumbria, NWS.
NWS (Nuclear Waste Services). n.d. b. *Introduction to Geological Disposal*. Calderbridge Cumbria, NWS.
NWS (Nuclear Waste Services). 2021. *GDF Annual Report 2020–2021*. Calderbridge Cumbria, NWS.
NWS (Nuclear Waste Services). 2022. *Geological Disposal Facility Creating Jobs & Skills: A First Look*. Calderbridge Cumbria, NWS.
Ocko, P. 2022. *Moonhaven*. New York, AMC.
OFGEM. 2019. *Ofgem Decarbonisation Programme Action Plan*. London, OFGEM.
Ogunrinola, G. A., Oyewale, J. O., Oshamika, O. O., & Olasehinde, G. I. 2020. The human microbiome and its impacts on health. *International Journal of Microbiology*, (1), 8045646.
Ojala, M., Cunsolo, A., Ogunbode, C. A., & Middleton, J. 2021. Anxiety, worry, and grief in a time of environmental and climate crisis: A narrative review. *Annual Reviews of Environment and Resources*, 46, 35–58.
Oliveira, R. V. 2022. Irreplaceable good: Bridging sustainability and intergenerational sufficientarianism. *Ethics, Policy and Environment*, 1–17.
O'Neill, D. W., Fanning, A. L., Lamb, W. F., & Steinberger, J. K. 2018. A good life for all within planetary boundaries. *Nature Sustainability*, 1, 88–95.
O'Riordan, T. 2000. Environmental science on the move. In: O'Riordan, T. (ed.) *Environmental Science for Environmental Management*. Abingdon, Routledge.

Oswald, Y., Owen, A., & Steinberger, J. K. 2020. Large inequality in international and intranational energy footprints between income groups and across consumption categories. *Nature Energy*, 5.

Oxfam. 2022. *Inequality Kills: The Unparalleled Action Needed to Combat Unprecedented Inequality in the Wake of COVID-19 – Summary.* Oxford, Oxfam International.

Page, E. A. 2007a. Fairness on the day after tomorrow: Justice, reciprocity and global climate change. *Political Studies*, 55(1), 225–242.

Page, E. A. 2007b. Justice between generations: Investigating a sufficientarian approach. *Journal of Global Ethics*, 3, 3–20.

Page, E. A. 2007c. Intergenerational justice of what: Welfare, resources or capability. *Environmental Politics*, 16, 453–469.

Paradies, Y. 2020. Unsettling truths: Modernity (de-)coloniality and Indigenous futures. *Postcolonial Studies*, 24, 438–456.

Parfit, D. 2000. Equality of priority? In: Clayton, M., & Williams, A. (eds.) *The Ideal of Equality*. Basingstoke, Palgrave.

Parfit, D. 2017. Future people, the non-identity problem, and person-affecting principles. *Philosophy & Public Affairs*, 45, 118–157.

Parrique, T., Barth, J., Briens, F., Kerschner, C., Kraus-Polk, A., Kuokkanen, A., & Spangenberg, J. H. 2019. *Decoupling Debunked. Evidence and Arguments Against Green Growth as a Sole Strategy for Sustainability*. A study edited by the European Environment Bureau EEB. [Online]. [Accessed March 2024]. https://www.almendron.com/tribuna/wp-content/uploads/2019/11/decoupling-debunked.pdf

Partridge, J. 2022. 'It's just not worth it': Why employers still can't get staff back to the office. *The Observer*.

Patel, R., & Moore, J. 2020. *A History of the World in Seven Cheap Things*. Berkeley, University of California Press.

Pearson, A. R., Tsai, C. G., & Clayton, S. 2021. Ethics, morality, and the psychology of climate justice. *Current Opinion in Psychology*, 42, 36–42.

Peck, J. 2010. *Constructions of Neoliberal Reason*. Oxford, Oxford University Press.

Pettifor, A. 2019. *The Case for the Green New Deal*. London, Verso.

Petzold, J., Andrews, N., Ford, J. D., Hedemann, C., & Postigo, J. C. 2020. Indigenous knowledge on climate change adaptation: A global evidence map of academic literature. *Environmental Research Letters*, 15(11), 113007.

Pien, C. 2022. Dawn of the space lords. *The Baffler*.

Piketty, T. 2013. *Capital in the Twenty-First Century*. Cambridge, MA, The Belknap Press of Harvard University Press.

Piketty, T. 2020. *Capital and Ideology*. Cambridge, MA, The Belknap Press of Harvard University.

Plato. 2000. *The Republic*. Cambridge, Cambridge University Press.

Plough, A., & Krimsky, S. 1987. The emergence of risk communication studies: Social and political context. *Science, Technology, & Human Values*, 12, 4–10.

Posiva. 2021. *Annual Report 2021*. Helsinki, Posiva.

Posiva. 2023. *Long-Term Safety*. [Online]. [Accessed December 2022]. https://www.posiva.fi/en/index/finaldisposal/long-termsafety.html

Powell, D. E. 2018. *Landscapes of Power: Politics of Energy in the Navajo Nation*. Durham, NC, Duke University Press.

Pratchett, T. 1991. *Witches Abroad*. London, Victor Gollancz.

Prăvălie, R. 2014. Nuclear weapons tests and environmental consequences: A global perspective. *Ambio*, 43(6), 729–744.

Price, J., Keppo, I., & Dodds, P. E. 2023. The role of new nuclear power in the UK's net-zero emissions energy system. *Energy*, 262, 125450.

Pyyhtien, O. 2016. *The Gift and its Paradoxes: Beyond Mauss*. Abingdon, Routledge.

Quijano, A. 2010. Coloniality and modernity/rationality. In: Mignolo, W. D., & Escobar, A. (eds.) *Globalisation and the Decolonial Option*. Abingdon, Routledge.

Quilter-Pinner, H., Statham, R., Jennings, W., & Valgarosson, V. 2021. *Trust Issues: Dealing with Distrust in Politics*. London, Institute of Public Policy Research.

Ramana, M. V. 2020. Why Technical solutions are insufficient: The abiding conundrum of nuclear waste. In: Brunnengräber, A. D., & Di Nucci, M. R. (eds.) *Conflicts, Participation and Acceptability in Nuclear Waste Governance: An International Comparison Volume III*. Heidelberg, Springer VS.

Rawls, J. 1993. The law of peoples. *Critical Enquiry*, 20, 36–68.

Rawls, J. 2009. *A Theory of Justice*. Cambridge, MA, Harvard University Press.

Raworth, K. 2022. *Doughnut Economics: Seven Ways to Think Like a 21st-Century Economist*. London, Chelsea Green Publishing.

RCEP (Royal Commission on Environmental Pollution). 1976. *Sixth Report: Nuclear Power and the Environment*. Cmnd 6618, HMSO. [Online]. [Accessed March 2024]. https://webarchive.nationalarchives.gov.uk/20110322144120/; http://www.rcep.org.uk/reports/06-nuclear/1976-06nuclear.pdf.

Robeyns. 2017. *Wellbeing, Freedom and Social Justice: The Capability Approach Re-Examined*. London, Open Book Publishers.

Robinson, J. 2006. *Economic Philosophy*. Abingdon, Routledge.

Robinson, J. B. 1982. Energy backcasting: A proposed method of policy analysis. *Energy Policy*, 10, 337–344.

Robinson, K. S. 1992. *Red Mars*. London, Spectra/Bantam Dell/Random House.

Robinson, K. S. 1993. *Green Mars*. London, Spectra/Bantam Dell/Random House.

Robinson, K. S. 1996. *Blue Mars*. London, Spectra/Bantam Dell/Random House.

Robinson, K. S. 2020. *The Ministry for the Future*. London, Orbit.

Roos, A. 2023. *Solar Technology and Global Environmental Justice: The Vision and the Reality*. Abingdon, Routledge.

Rosa, E. A., & Freudenburg, W. R. 1993. The historical development of public reactions to nuclear power: Implications for nuclear waste policy. In: Dunlap, R. E., Kraft, M. E., & Rosa, E. A. (eds.) *Public Reactions to Nuclear Waste:*

Citizens' Views of Repository Siting (pp. 32–63). Durham, NC, Duke University Press.

Ruff, T. A. 2015. The humanitarian impact and implications of nuclear test explosions in the Pacific region. *International Review of the Red Cross*, 97(899), 775–813.

Rushcoff, D. 2022. *Survival of the Richest: Escape Fantasies of the Tech Billionaires*. London, Scribe.

Sachs, J. 2014. Climate change and intergenerational wellbeing. In: Bernard, L., & Semmler, W. (eds.) *The Oxford Handbook of the Macroeconomics of Global Warming*. Oxford, Oxford University Press.

Saddington, K., & Templeton, W. L. 1958. *Disposal of Radioactive Waste*. London, George Newnes.

Sahlins, M. 2017. *Stone Age Economics*. London and New York, Routledge.

Saini, A. 2023. *The Patriarchs: How Men Came to Rule: Origins of Inequality*. London, HarperCollins Publishers.

Santos, B. S. 2016. Epistemologies of the South and the future. *From the European South*, 1, 17–29.

Schlosberg, D. 2004. Reconceiving environmental justice: Global movements and political theories. *Environmental Politics*, 13, 517–540.

Schlosberg, D. 2013. Theorising environmental justice: The expanding sphere of a discourse. *Environmental Politics*, 22, 37–55.

Schlosberg, J., & Laker, L. 2018. *Labour, Antisemitism and the News. A Disinformation Paradigm*. London, Media Reform Coalition.

Scott, J, C. 1999. *Seeing Like a State: How Certain Schemes to Improve the Human Condition Have Failed*. Ithaca, Yale University Press.

Sen, A. 2009. *The Idea of Justice*. Cambridge, MA, The Bellnap Press.

Shahen, M., Kotani, K., & Saijo, T. 2021. Intergenerational sustainability is enhanced by taking the perspective of future generations. *Scientific Reports*, 11, 2437.

Shields, L. 2020. Sufficientarianism. *Philosophy Compass*, 15(11), 1–10.

Shiva, V. 2007. Biodiversity, intellectual property rights, and globalisation. In: Santos, B. D. S. (ed.) *Another Knowledge Is Possible: Beyond Northern Epistemologies*. London, Verso.

Shue, H. 2014. *Climate Justice: Vulnerability and Protection*. Oxford, Oxford University Press.

Simard, S. W. 2018. Mycorrhizal networks facilitate tree communication, learning, and memory. In: Baluska, F., Gagliano, M., & Guenther, W. (eds.) *Memory and Learning in Plants*. Berlin, Springer.

Simis, M. J., Madden, H., Cacciatore, M. A., & Yeo, S. K. 2016. The lure of rationality: Why does the deficit model persist in science communication? *Public Understanding of Science*, 25, 400–414.

Simpson, J., Inman, P., & Adu, A. 2024. Labour has no intention of reinstating cap on bankers' bonuses, says Reeves. *The Guardian*. [Online]. [Accessed March 2024]. https://www.theguardian.com/business/2024/jan/31/labour-party-cap-bankers-bonuses-financial-services

Sioui, G. E. 1999. *Huron Wendat: The Heritage of the Circle*. Vancouver, UBC Press.

Skrimshire, S. 2018. Confessing Anthropocene. *Environmental Humanities*, 10, 310–329.

Smith, L. T. 1999. *Decolonising Methodologies: Research and Indigenous Peoples.* London and New York, Zed Books Ltd.

Smith, M. A. 2016. The UK and nuclear weapons. In: Brown, D (ed.) *The Development of British Defence Policy: Blair, Brown and Beyond* (pp. 169–188). Routledge, Abingdon.

Smith, M. L., & Seward, C. 2009. The relational ontology of Amartya Sen's capability approach: Incorporating social and individual causes. *Journal of Human Development and Capabilities*, 10, 213–235.

Smyth, D. 2012. *Response to West Cumbria MRWS Consultation: Why a Deep Nuclear Waste Repository Should Not Be Sited in Cumbria.* [Online]. [Accessed December 2023]. https://www.davidsmythe.org/nuclear/Smythe%20consultation%20response%20march%202012.pdf

Solnit, R. 2004. *Hope in the Dark: The Untold History of People Power.* London, Canongate Books.

Soper, K. 2020. *Post-Growth Living: For an Alternative Hedonism.* London, Verso Books.

Sovacool, B. K. 2008. Valuing the greenhouse gas emissions from nuclear power: A critical survey. *Energy Policy*, 36, 2950–2963.

Sovacool, B. K. 2019. The precarious political economy of cobalt: Balancing prosperity, poverty, and brutality in artisanal and industrial mining in the Democratic Republic of the Congo. *The Extractive Industries and Society*, 6(3), 915–939.

Sovacool, B. K., Baum, C. M., & Low, S. 2023. Beyond climate stabilization: Exploring the perceived sociotechnical co-impacts of carbon removal and solar geoengineering. *Ecological Economics*, 204, 107648.

Sovacool, B. K., Burke, M., Baker, L., Kotikalapudi, C. K., & Wlokas, H. 2017. New frontiers and conceptual frameworks for energy justice. *Energy Policy*, 105, 677–691.

Sovacool, B. K., & Dunlap, A. 2022. Anarchy, war, or revolt? Radical perspectives for climate protection, insurgency and civil disobedience in a low-carbon era. *Energy Research & Social Science*, 86, 102416.

Sovacool, B. K., Hook, A., Martiskainen, M., Brock, A., & Turnheim, B. 2020. The decarbonisation divide: Contextualizing landscapes of low-carbon exploitation and toxicity in Africa. *Global Environmental Change*, 60, 102028.

Spannring, R. 2021. Youth in the Anthropocene: Questions of intergenerational justice and learning in a more-than-human world. In: Knapp, G., & Krall, H. (eds.) *Youth Cultures in a Globalized World Developments, Analyses and Perspectives*. Berlin, Springer.

Sperber, E. 2018. The emergency brake. *Counterpunch.*

Spinardi, G. 1997. Aldermaston and British nuclear weapons development: Testing the 'Zuckerman thesis'. *Social Studies of Science*, 27, 547–582.

Spinoza, B. D. 1982. *The Ethics and Selected Letters*. Indianapolis, Hackett.

Stacey, K. 2024. Starmer to scale back home insulation scheme as part of green policy U-turn. *The Guardian.* [Online]. [Accessed March 2024]. https://www.theguardian.com/politics/2024/feb/08/starmer-to-scale-back-home-insulation-scheme-as-part-of-green-policy-u-turn

Stacey, K., & Crerar, P. 2024. Starmer to announce scaling back of £28bn green investment plan. *The Guardian.* [Online]. [Accessed March

2024]. https://www.theguardian.com/politics/2024/feb/07/keir-starmer-to-announce-scaling-back-of-28bn-green-investment-plan
Standard Eurobarometer. 2017. *Public Opinion in the European Union.* Brussels, European Commission.
Standard Eurobarometer. 2022. *Europeans' Opinions about the European Union's Priorities*. Brussels, European Commission.
Stanley, A. 2009. Just space or spatial justice? Difference, discourse, and environmental justice. *Local Environment*, 14, 999–1014.
Statista. 2023. *Value of Quantitative Easing Measures by the Bank of England (BoE) in the United Kingdom from November 2009 to November 2020*. Statista. [Online]. [Accessed June 2023]. https://www.statista.com/statistics/1105570/value-of-quantitative-easing-by-the-bank-of-england-in-the-united-kingdom/
Steffen, W., Grinevald, J., Crutzen, P., & Mcneill, J. 2011. The Anthropocene: Conceptual and historical perspectives. *Philosophical Transactions of the Royal Society A: Mathematical, Physical and Engineering Sciences*, 396, 842–867.
Stern, N. 2007. *The Economics of Climate Change: The Stern Review*. Cambridge, Cambridge University Press.
Stirling, A. 2007. Risk, precaution and science: Towards a more constructive policy debate. *EMBO (European Molecular Biology Organisation) Report*, 8(4).
Stirling, A. 2015. *Towards Innovation Democracy? Participation, Responsibility and Precaution in Innovation Governance: Innovation Democracy.* Brighton, STEPS Centre, University of Sussex.
Stirling, A., & Johnstone, P. 2018. *A Global Picture of Industrial Interdependencies Between Civil and Military Nuclear Infrastructures.* Brighton, SPRU Science and Policy Unit, University of Sussex.
Stoddard, I., Anderson, K., Capstick, S., Carton, W., Depledge, J., Facer, K., Gough, C., Hache, F., Hoolohan, C., Hultman, M., & Hällström, N. 2023. Three decades of climate mitigation: Why haven't we bent the global emissions curve? *Annual Review of Environment and Resources*, 46, 653–689.
STUK. 2018a. *The Radiation and Nuclear Safety Authority's Strategy for 2018*–2022. Helsinki, STUK.
STUK. 2018b. *Disposal of Nuclear Waste (Guide YVL D.5)*. Helsinki, STUK.
STUK. 2022. *Finnish Report on Nuclear Safety: Finnish 9th National Report as Referred to in Article 5 of the Convention on Nuclear Safety.* Helsinki, STUK.
Sturgis, P., & Allum, N. 2004. Science in society: Re-evaluating the deficit model of public attitudes. *Public Understanding of Science*, 13, 55–74.
Suldovsky, B. 2016. In science communication, why does the idea of the public deficit always return? Exploring key influences. *Public Understanding of Science*, 25, 415–426.
Szulecki, K. 2018. Conceptualizing energy democracy. *Environmental Politics*, 27, 21–41.
Taebi, B., Roeser, S., & Van De Poel, I. 2021. The ethics of nuclear power: Social experiments, intergenerational justice, and emotions. *Energy Policy*, 51, 202–206.

Táíwò, O. O. 2022. *Reconsidering Reparations.* Oxford, Oxford University Press.
Taylor, A. 2019. *Democracy May Not Exist but We'll Miss It When It's Gone*. London, Henri Holt & Co.
Taylor, A. 2021. *Remake the World: Essays, Reflections, Rebellions.* Chicago, Haymarket Books.
Teräväinen, T., Lehtonen, M., & Martiskainen, M. 2011. Climate change, energy security, and risk – debating nuclear new build in Finland, France and the UK. *Energy Policy*, 39, 3434–3442.
The Bulletin of Atomic Scientists. 2024. A moment of historic danger: It is still 90 seconds to midnight. *The Bulletin of Atomic Scientists*. [Online]. [Accessed January 2024]. https://thebulletin.org/doomsday-clock/
The Equality Trust. 2022. Equality trust finds a 1000% increase in billionaire wealth. *The Equality Trust*. [Online]. [Accessed January 2024]. https://equalitytrust.org.uk/news/equality-trust-finds-1000-increase-billionaire-wealth
Theddlethorpe GDF Community Partnership. 2024a. *Resources (Minutes to Working Group and Community Partnership Meetings)*. [Online]. [Accessed March 2024]. https://theddlethorpe.workinginpartnership.org.uk/resources/
Theddlethorpe GDF Community Partnership. 2024b. *Contact Us*. [Online]. [Accessed March 2024]. https://theddlethorpe.workinginpartnership.org.uk/contact-us/
Thomas, S. 2023. UK nuclear waste policy: 50 wasted years. In: Arentsen, M., & Van Est, R. (eds.) *The Future of Radioactive Waste Governance: Lessons from Europe* (pp. 199–229). Berlin, Springer.
Thompson, D. F. 2010. Representing future generations: Political presentism and democratic trusteeship. *Critical Review of International Social and Political Philosophy*, 13, 17–37.
Titans of Nuclear. 2022, August 16. *David Sterns – Nuclear Financing Partner*. [Podcast], Titans of Nuclear. https://www.lastenergy.com/titansofnuclear/experts/davidstearns
Tol, R. S. J. 2009. The economic effects of climate change. *The Journal of Economic Perspectives*, 23(2), 29–51.
Trischler, H., & Bud, R. 2018. Public technology: Nuclear energy in Europe. *History and Technology*, 34(3–4), 187–212.
UK Government Investments. 2023, November. *Annual Report on the UK Government's Contingent Liabilities*. [Online]. [Accessed January 2024]. file:///C:/Users/Leopold/Downloads/CLCC-Annual-Report-on-the-UK-Governments-Contingent-Liabilities-November-2023-Web-Accessible.pdf
UNICEF. 2023. *Child Poverty in the Midst of Wealth.* Innocenti Report Card, UNICEF. [Online]. [Accessed January 2024]. https://www.unicef.org/globalinsight/media/3291/file/UNICEF-Innocenti-Report-Card-18-Child-Poverty-Amidst-Wealth-2023.pdf
Van De Graaf, T., & Sovacool, B. K. 2020. *Global Energy Politics*, London, John Wiley & Sons.
Van Dijk, N. 2021a. From exacerbating the Anthropocene's problems to intergenerational justice: An analysis of the communication procedure of the human rights treaty system. *Earth System Governance*, 10, 100123.

Van Dijk, N. 2021b. The capability approach as a road map for re-thinking intergenerational justice. In: Linehan, J., & Lawrence, P. (eds.) *Giving Future Generations a Voice Normative Frameworks, Institutions and Practice*. Cheltenham, Edward Elgar Publishing.

Van Dyke, J. M. 1988. Ocean disposal of nuclear wastes. *Marine Policy*, 12(2), 82–95.

Van Uffelen, N., Taebi, B., & Pesch, U. 2024. Revisiting the energy justice framework: Doing justice to normative uncertainties. *Renewable and Sustainable Energy Reviews*, 189, 113974.

Van Wyck, P. C. 2010. *The Highway of the Atom.* London, McGill-Queen's University Press.

Vergès, F. 2019. *A Decolonial Feminism*. London, Pluto Press.

Vergès, F. 2020. *A Feminist Theory of Violence*. London, Pluto Press.

Verma, A. 2021. The nuclear, the humanities, and social science nexus: Challenges and opportunities for speaking across the disciplinary divides. *Nuclear Technology*, 207.

Vettese, T., & Pendergrass, D. 2022. *Half Earth Socialism: A Plan to Save the Future from Extinction, Climate Change and Pandemics*. London, Verso.

Vilhunen, T., Kojo, M., Litmanen, T., & Taebi, B. 2022. Perceptions of justice influencing community acceptance of spent nuclear fuel disposal. A case study in two Finnish nuclear communities. *Journal of Risk Research*, 25, 1023–1046.

Von Hippel, F. N. 2021. Plutonium programs in East Asia and Idaho will challenge the Biden administration. *Bulletin of the Atomic Scientists*. [Online]. [Accessed June 2023]. https://thebulletin.org/2021/04/plutonium-programs-in-east-asia-and-idaho-will-challenge-the-biden-administration/#post-heading

Voyles, T. B. 2015. *Wastelanding: Legacies of Uranium Mining in Navajo Country*. Minneapolis and St Paul, University of Minnesota Press.

Walia, H. 2021. *Border and Rule: Global Migration, Capitalism, and the Rise of Racist Nationalism*. Chicago, Haymarket Books.

Walker, C., & Walker, A. 2022. *Nuclear Energy in the UK*. London, UK Parliament POST (POSTnote 687).

Walker, G. 2009. Beyond distribution and proximity: Exploring the multiple spatialities of environmental justice. *Antipode*, 41, 614–636.

Wallerstein, I. M. 2006. *World-Systems Analysis: An Introduction*, Durham, NC, Duke University Press.

Wanlin, M., & Tiura, S. 2006. *Developing Effective Two-Way Communications Between Canada's Aboriginal Community and the NWMO.* Vancouver, Wanlin & Co.

Ward, L., Fleisher, S., & Towers, L. 2021. *Understanding Fairness Between Different Generations in Times of COVID-19*. University of Brighton Research Report. Brighton, University of Brighton.

Watane, K. 2022. *Reimagining the Human-Environment Relationship: Indigenous Philosophy and Intergenerational Justice.* Tokyo, United Nations University – Centre for Policy Research.

Watkins, S. 2022. Editorial: Five wars in one: The battle for Ukraine. *New Left Review*, 137, 5–20.

WCMRWSP (West Cumbria Managing Radioactive Waste Safely Partnership). 2012. *The Final Report of the West Cumbria Managing Radioactive Waste Safely Partnership.* Whitehaven, Copeland Borough Council.

Welsh, I. 1993. The NIMBY syndrome: It's significance in the history of the nuclear debate in Britain. *British Journal of the History of Science*, 26, 15–32.

Wheatley, S., Sovacool, B. K., & Sornette, D. 2017. Of disasters and dragon kings: A statistical analysis of nuclear power incidents and accidents. *Risk Analysis*, 37, 99–115.

Whyte, K., P. 2017. Is it colonial déjà vu? Indigenous peoples and climate injustice. In: Adamson, J., & Davis, M. (eds.) *Humanities for the Environment: Integrating Knowledge, Forging New Constellations of Practice*. Abingdon, Routledge.

Williams, R. 1989. *Resources of Hope: Culture, Democracy, Socialism.* London, Verso.

Winner, L. 1980. Do artifacts have politics? *Daedalus*, 121–136.

Winter, C. J. 2022. *Subjects of Intergenerational Justice: Indigenous Philosophy, the Environment and Relationships*. London, New York, and Abingdon, Routledge.

WNA (World Nuclear Association). 2023a. *Supply of Uranium*. [Online]. [Accessed June 2023]. https://world-nuclear.org/information-library/nuclear-fuel-cycle/uranium-resources/supply-of-uranium.aspx

WNA (World Nuclear Association). 2023b. *World Uranium Mining Production*. [Online]. [Accessed June 2023]. https://world-nuclear.org/information-library/nuclear-fuel-cycle/mining-of-uranium/world-uranium-mining-production.aspx

WNA (World Nuclear Association). 2024. *Nuclear Power in Canada*. [Online]. [Accessed June 2023]. https://world-nuclear.org/information-library/country-profiles/countries-a-f/canada-nuclear-power.aspx

World Bank. 2020. *Poverty and Inequality Platform*. China. [Online]. [Accessed June 2023]. https://pip.worldbank.org/country-profiles/CHN

Wynne, B. 1989. Sheepfarming after Chernobyl: A case study in communicating scientific information. *Environment: Science and Policy for Sustainable Development*, 31, 10–39.

Wynne, B. 1992. Uncertainty and environmental learning: Reconceiving science and policy in the preventive paradigm. *Global Environmental Change*, 2(2), 111–127.

Wynne, B. 2006. Public engagement as a means of restoring public trust in science–hitting the notes, but missing the music? *Public Health Genomics*, 9, 211–220.

Wynter, S. 2003. Unsettling the coloniality of being/power/truth/freedom: Towards the human, after man, its overrepresentation – an argument. *The New Centennial Review*, 3, 257–337.

Yagami, K. 2009. Bombing Hiroshima and Nagasaki: Gar Alperovitz and his critics. *Southeast Review of Asian Studies*, 31, 301–307.

Young, I. M. 1990. *Justice and the Politics of Difference.* Princeton, NJ, Princeton University Press.

Young, I. M. 1996. Communication and the other: Beyond deliberative democracy. In: Benhabib, S. (ed.) *Democracy and Difference: Contesting the Boundaries of the Political*. Princeton, NJ, Princeton University Press.

Young, S. 2024. *Why the Biden Administration's New Nuclear Gravity Bomb Is Tragic*. [Online]. [Accessed June 2023]. https://thebulletin.org/2024/02/why-the-biden-administrations-new-nuclear-gravity-bomb-is-tragic/?utm_source=Newsletter&utm_medium=Email&utm_campaign=ThursdayNewsletter02152024&utm_content=NuclearRisk_GravityBomb_02132024

Zabern, L. V., & Tulloch, C. D. 2020. Rebel with a cause: The framing of climate change and intergenerational justice in the German press treatment of the Fridays for Future protests. *Media, Culture & Society*, 43, 23–47.

Index

For Product Safety Concerns and Information please contact our EU representative GPSR@taylorandfrancis.com
Taylor & Francis Verlag GmbH, Kaufingerstraße 24, 80331 München, Germany

www.ingramcontent.com/pod-product-compliance
Ingram Content Group UK Ltd.
Pitfield, Milton Keynes, MK11 3LW, UK
UKHW022344190926
475702UK00015B/2096

* 9 7 8 1 0 3 2 7 2 8 0 0 1 *